The *Theory* of Extuity

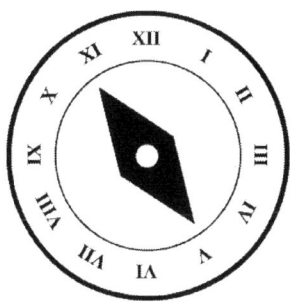

A reality-changing **new theory of 'time'** that unveils humanity's greatest ability.

Hakeem Javaid

Copyright © 2024 Hakeem Javaid

The right of Hakeem Javaid to be identified as the author of this work has been asserted by him in accordance with the Copyright, Designs and Patents Act 1988.

No part of this publication may be reproduced, or stored in a retrieval system, or transmitted in any form or by any means, electronic, mechanical, photocopying, recording, or otherwise, without express written permission of the publisher.

The information in this book is not intended to replace the advice of the reader's own medical professional. You should consult a medical professional in matters relating to health, including mental health. Individual readers are solely responsible for their own health care decisions. The author does not accept responsibility for any adverse effects individuals may claim to experience, whether directly or indirectly, from the information contained in this book.

First Edition: 2024

"Amazing. Epic. Groundbreaking. I absolutely loved it. There was a part where it just totally made sense. I genuinely think this could be HUGE. So exciting... so so exciting."

Katie Stoddart

International Keynote Speaker, Award-Winning Leadership Coach, TEDx Host, and Author.

"As you're reading, you can tell that the author has broken through a barrier nobody else has yet... it's almost like you wake up to what's going on."

Jo Reid

Celebrity Mindset Coach, Biofield Frequency Therapist, Cellular Health Speaker, and Author.

"Oh my gosh, so much happening right now... it makes you rethink everything you thought you knew about how life works. Dammit... my brain won't shut off."

Jenny Thrasher

Founder of Thrive Education, Mental Health Consultant, International Speaker, and Author.

"This book will not only open your eyes to new concepts of the mind and of self but will introduce you to how we can in fact influence time and are actually already doing so without even realising it. In time, you will understand that reading this book was a turning point in your life, but maybe you already have..."

Gemma Walker
Certified Neuro-Linguistic Programming (NLP) Practitioner,
Certified Hypnotherapist, and Author.

"The Matrix is real."

Anonymous
Engineer at a Military Weapons Manufacturer

"Truly mind-blowing and eye-opening! The way Hakeem was able to relate extreme concepts into easily understandable bits of knowledge really accelerated the understanding of what was being discussed. Extuity makes complete sense."

Rich Ranieri
International English Language Consultant, Chi Practitioner,
and Non-Ordinary States Specialist.

Contents

Introduction	1
Level 1 • Seriously, WTF is 'Time'?	17
Side-Level: Nature's Clocks	41
Level 2 • Time travel is obviously… wait, what?	49
Side-Level: Out-of-Body	73
Level 3 • Quantum ~~Physics~~ Biology	79
Side-Level: Our Quantum Bodies	101
Level 4 • The Word With A Bullsh*t Definition	105
Side-Level: Mind, Body, Soul?	121
Level 5 • The Theory of Extuity	127
Side-Level: A Time Hack	143
Level 6 • The Language of Extuity	149
Side-Level: Past Lives?	179
Level 7 • Humanity's Greatest Secret	183
Side-Level: A Final Gift	217
Final Words	223
Notes	241

Introduction

A Journey of Impossibility

There are very few discoveries in known history that have redefined humanity's understanding of 'reality'. Or rather, discoveries that have caused such a dramatic shift in reality that meant humanity simply couldn't look at life the same way again. In 350 BC, the Theory of a Spherical Earth. In 1666, the Theory of Gravity. In 1905, the Theory of Relativity.

Whether the Theory of Extuity is one of those paradigm shifts, I'll leave that for you to decide.

However, since I discovered it in 2023, I truly have not seen reality in the same way. I cannot. And the people I've shared it with have *all* reacted in the same way, *every single one*. Their minds have raced a million miles an hour as they've tried to comprehend infinite new possibilities.

And I really do mean *infinite*.

To this day, I still haven't uncovered the full implications of it.

The Theory of Extuity proposes a fundamental change to our reality because it completely redefines humanity's relationship with time. As you'll soon discover, 'extuity' is specifically a human ability that we've been unconscious of, one deeply intertwined with time, and one that reveals entirely new answers to our greatest questions and inexplicable phenomena: coincidence, deja vu, manifestation, the Mandela Effect, the law of attraction, and, yes, even *fate*.

It's a *reality-changing* revelation, one there is no shortcut to understanding, hence this book.

This journey will most probably be unlike anything you have ever or will ever experience, primarily because it will challenge your deepest beliefs of reality, your very understanding of the world around you *and* within you. Therefore, more important than anything is the mindset you take with you on this journey.

So, we're going to start with my unique story, a journey of crazy impossibilities.

Not only will this help cultivate the right mindset for this book, but I'd like you to get to know me a little better to understand the values and principles that led to discovering this theory. Many also find my story inspiring, so hopefully it will inspire your application of the Theory of Extuity to build that impossible life you've always envisioned.

And if you're reading this book for a second time, this time with an understanding of this theory, you're about to read my story in a completely different light. You'll see revelations hidden in plain sight, ones that I myself only saw once I understood extuity.

So... are you ready?
Perfect.
Let's dive in.

Navigating Impossibility

There are three key points in time where I've faced decisions that felt impossible to answer. Each time I was heavily encouraged to choose a path that every fibre of my being told me wasn't right for me, and each time I made the extremely difficult choice that shocked everyone around me. I just 'knew' that each choice led to a better future. And guess what? They did. They took me on a journey that many believed impossible.

The first was back in 2012.

I managed to receive offers from every university I applied to, which included the University of Cambridge, one of the top three universities in the world. Great, right? Well… not if you didn't want to go, and, yes, I didn't wish to.

If your first thought is that I might have had good reason for not wishing to, I love you already. Cherish your open mind. If, however, you think that's crazy, you're not alone. Many did back then. It goes against the social norm. However, please embrace your open mind, it's absolutely critical to understanding the implications of what I share in this book. To even begin to consider a new reality, we must be prepared to let go of everything we think we know.

So, let's hear the full story…

What few people know is that Imperial College London, another top-tier UK university, was a better university for the specific course I wished to study: Computer Science. In fact, they call it 'Computing' because it's so much more practical than any other university. Not only were Imperial computing graduates the most sought-after computer scientists in the UK, but more importantly, I'm someone who learns best from practical experience. Imperial was by far the best option.

Not everyone agreed though.

My parents were so adamant I was going to Cambridge, they told all our relatives I'd accepted the offer. When one of my relatives brought up the topic, they said I'd be crazy to turn down Cambridge. I replied, "It doesn't matter what university you go to, what matters is what you learn, and more importantly, what you do with what you learn."

They didn't bring up the topic again.

What our egos constantly blind us to is the fact that labels and prestige mean nothing. People's opinions are just people's opinions. What matters in life is what you experience, what you learn, and what you do with that. That gave Imperial more bonus points because I wished to start my own tech company once I mastered computing, and building up my professional network in London while studying was a huge bonus.

So, I had a decision to make. Make my parents and relatives proud and raise my social status, or let them think I'm crazy and follow a path I believed would be better. In the end, I remember asking myself: "50 years from now, which decision would I regret?", and I had my answer. I turned down the offer to study the more theoretical course at the University of Cambridge, and I accepted the offer to study the more practical course at Imperial College London.

Funnily enough, the second impossible decision I faced came soon after.

In 2014, I chose to drop out of university.

Crazy? Maybe, but let's keep that open mind.

It turned out that Imperial's computing course was far better than I expected. I learnt how to code every major language and work with every major technology system in just one year. The 2nd and 3rd years of the course were highly project-based, and I'll be honest with you, I absolutely hated putting my time and effort into work just to be graded. I could be putting that time into real-world projects helping real people.

Which, by the way, I'd already experienced.

I actually started coding at the age of 11, began building websites for local businesses at 14, and right after my first year of university, I started a hardware company with three friends, which exploded. In a good way.

The video I produced for it went viral. It was picked up by major tech publications and we were being approached by investors worldwide. We won awards from Intel, the Mayor of London, and even the Royal Family. My parents were thrilled. The mental scars of their son turning down Cambridge were long forgotten. My university, Imperial College London, was also supporting us all the way.

And that's what made this decision so difficult.

Do I remain studying at university, sacrificing my time for grades and a certificate that could get me into the biggest tech companies in the world? Or do I drop out and work full time on a company that I was building with my own two hands, with experience that had already taught me far more than my first year of university already had?

Of course, everyone was telling me to stay at university, to get the degree, to secure that safety net. But, if you really knew me, I never had any intention of working for a company. I had every intention of building and working on my own ventures. So, it was a decision between following society's safe route, or the risky option of forging my own path.

In the end, I followed what I believed was right for me. I just knew my future self would regret it if I didn't. So, as difficult as it was, I dropped out of university in my second year and I worked full-time on building my future.

Oh, and guess what... a few months later, we were told by leading entrepreneurs around London that the tech company we were building was worth over £3 million, which, of course, led to the third impossible decision.

In 2015, a year later, I decided to leave that tech company.

If you're thinking there must have been a good reason for me to do so, please take a moment to appreciate your open mind, and do your best to maintain that mindset throughout this book. It will take you further than you can imagine in life, and it's absolutely critical for the mind-bending journey ahead.

So... this third decision, the one that led to months of depression. Was there a reason I made such a difficult call?

Yes, most certainly so.

The tech company in question was a hardware company building a revolutionary smartwatch at a time when the market was young. Apple and Samsung hadn't entered with their own products yet, but we heard from behind the scenes they had every intention to. So, what did that mean for us? Nothing good.

More impactful, however, was what came with being told the valuation of our company. Over £3 million. I was only 20 years old! My three co-founders themselves were only in their mid-twenties. Unfortunately, as we are often warned in life, people 'change' when it comes to money. I've since learnt that people don't actually change. Instead, their true natures are revealed.

I discovered that one of my three co-founders was someone that I had no wish to work with at all. Once the £3 million valuation was revealed, I saw their true nature. I'd always felt that there was some disconnect between us, and now it was as clear as day.

So, I faced an impossible decision…

Do I spend the next five to ten years working on a hardware company that I wouldn't enjoy building, but with a potentially big paycheck at the end? (If Apple, Google or Samsung didn't take over the market first). Or do I leave, use my skills in building software companies where my true passion was, and live a life I'd potentially enjoy significantly more?

It was effectively a choice of money vs happiness. It's a question you ask friends to casually pass the time, yet it was one I had to make that would decide where my life would lead. So, I asked myself once again: "50 years from now, which decision would I regret?"

You of course know what decision I made, and it crippled me. It was like leaving behind a child you'd raised. I'd put so much emotional energy, time, blood, sweat and tears into building the company. It taught me so much about the real world. It helped me grow more than I ever expected.

But, I had to trust my gut. I had to trust myself.

And, oh my, am I glad I did.

The Impossible Timeline

Nearly 10 years later, I write this, aged 29, sitting on the coast of Madeira Island, having left the UK's cold winter behind, spending my days working on multiple tech companies and hiking in some of the most beautiful landscapes on Earth.

I'm living a life that many dream of, a life that, I was astonished to realise, may not have been built by just my past self, but by my future self too.

Yes, you read that right.

With what I've discovered, with the reality-shifting new theory that my highly logical yet creative mind keeps arriving at, it turns out that my future self may have partially written my past and present. With what I've uncovered, my future self may have been entirely responsible for the outcome of all those impossible decisions I faced.

Yes, it sounds absolutely crazy, and even though you are keeping a beautiful open mind, a crazy idea is what we want. Why? Because that's exactly what new discoveries start out as.

Crazy. Ridiculous. Incomprehensible.

And what's even crazier is where this all stems from.

2023.

2023 was a year that was not meant to be. It was a year that redirected me away from some of the biggest business plans I'd ever designed, throwing me on a path of life-changing, mind-blowing experiences. How? By kickstarting the year breaking my arm snowboarding in the French Alps.

A deep blessing in a very painful disguise.

I was rushed down the mountain for surgery, gifted with over a month of excruciatingly painful nights, and left on the verge of experiencing depression for the second time in my life.

I'm a rock climber, programmer, artist. My life as I knew it was torn away from me. It was out of a pure will to survive that my mind refused to let a 6-month recovery period being physically handicapped prevent me from growing and experiencing the wonders of the world, which left only one real option: redirecting my energy into growing mentally and diving into the mysteries of the mind.

Reminding myself to keep as wide an open mind as possible, frequently repeating the humbling quote "You don't know what you don't know", my year took me on a journey I'd never thought I'd experience. My journey into the mind led to understanding and consciously experiencing the different brainwave states, meditating with monks and self-inducing lucid dreams, feeling and directing the flow of Chi in my body, and challenging my views on every conditioned 'fact' I've been programmed with.

And this journey that we are about to embark on together started right there, when I began questioning pre-conditioned 'facts' in our minds, the lens through which we see reality. Specifically, our perception and understanding of 'time'; a seemingly directional concept, but in reality, a far more complicated and powerful puzzle than we could ever have imagined.

Diving between many of life's unanswered questions, my journey through the mysteries of time has led me to connecting the dots between some of the most confusing discoveries in Quantum

INTRODUCTION

Physics and some of the biggest inexplicable human experiences that we all encounter day to day.

One of the most unique things about me is that I possess both a highly logical mind and an extremely creative one too. I'm a software engineer, data scientist, and UI developer, yet also a graphic designer, video producer, and UX designer. And not just a 'little' experienced in each, I'm the outlier that's mastered each one. I've coded complex algorithms yet designed the most beautifully simple apps used by millions, winning numerous awards on the way. I believe it is this balance of logic vs creative out-of-the-box thinking that led to a revelation of time that left me in awe.

And that, my friend, is why we are both here.

The Theory of Extuity.

A Journey Through Time

The seven chapters you are about to read are no ordinary chapters. The seven chapters are in fact seven levels, each representing different levels of understanding of 'time'. Every level will open your mind to new understandings of time, each with specific revelations required to understand the next.

We begin on Level 1 by addressing the human construct of 'time', removing the fog of illusion. Level 2 dives into phenomena of time discovered by physicists, opening your mind to seemingly-impossible realities of time. Levels 3 and 4 dive into ancient monk teachings tied to phenomena of time that we experience as humans, opening your mind to seemingly-impossible human abilities. Level 5 onwards contain the reality-changing revelations of extuity.

And you'll find a little surprise at the end of each level.

With every new level of understanding of time comes a new perspective of reality. For each one, I discovered a new smaller theory addressing a mystery of our reality related to that understanding of time, most of which I haven't found mentioned elsewhere. I've included these as optional "side-levels".

As we rise through the levels, we make our way to the main event.

This new theory weaves together the powers of time and energy, explaining an accepted mystery of humanity whilst revealing infinite possibilities.

The Theory of Extuity.

For those who understand the journey we're about to embark on and its infinite implications, it won't just change your life going forward in time, but it will have already changed your past. In other words, your future self, knowing and applying what you're about to learn, may have partially written your past and present.

Yes, it's crazy, ridiculous, incomprehensible.

Perfect.

Henry Ford once said that if he asked what people wanted, they would have said faster horses. Henry Ford invented the car. It's a beautiful reminder that if you want to experience your greatest potential, you must dive into the unknown. Forgo all notion of 'faster horses'.

To accelerate beyond your current reality, you must be prepared to let go of it.

If I am right, the Theory of Extuity doesn't just answer the greatest phenomena that we experience day-to-day, but will also allow you to consciously navigate through the dimensions of time, past and future, to give you unprecedented control over your present. And the funny thing is, if it really is as powerful as I believe it to be, your future self may have been the one who decided for you to read this book.

So, my dear friend, if that sounds too much for you, know that there's absolutely nothing wrong with sticking with faster horses.

But, if you're with me, let's go on a journey together. A journey of wondrous possibilities.

A journey through time.

YOUR JOURNEY
Level Progress & Key Learning Points

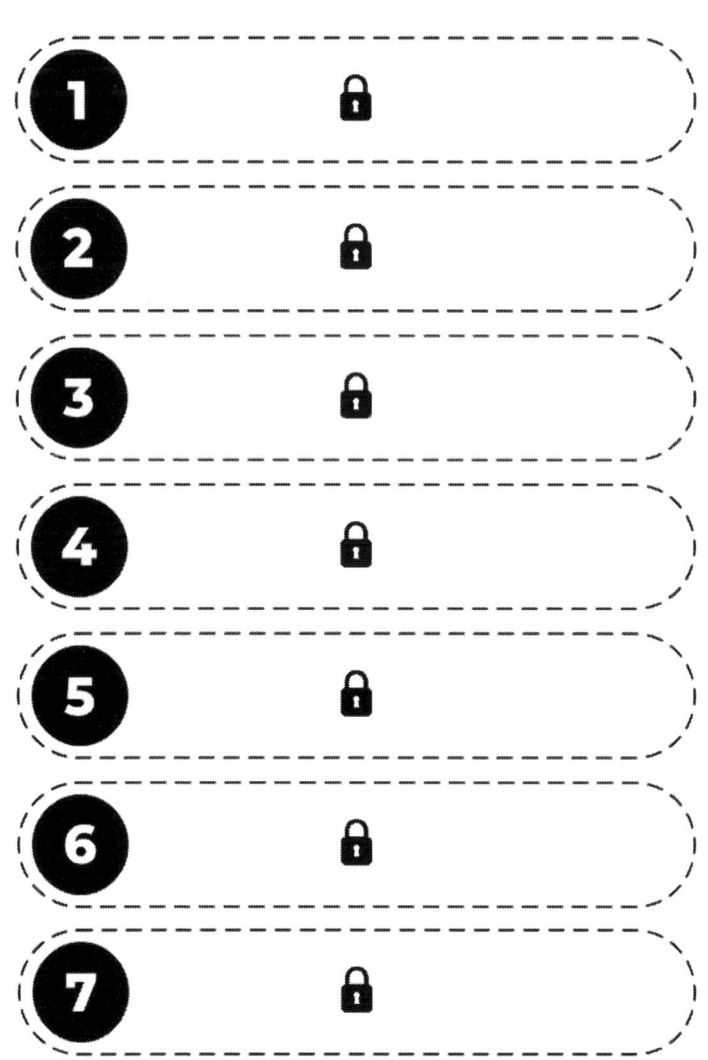

Level 1

Seriously, WTF is 'Time'?

In Physics, 'time' is a dimension that measures the sequence and duration of events in the universe. Psychology states it's the *perception* of the passage of events, regulated by the brain's internal clock. Poets describe it as an invisible river that carries all things from the past into the future. And a kid would probably tell you it's how long you can play for until bedtime.

To this day, there isn't a universally accepted definition of 'time'. That's why people say, "Time is an illusion".

Yet, many forget that illusions only exist because of a lack of understanding, and, yes, this does indeed apply to 'time'. Humanity, to this day, does *not* understand time. Think about that for a second. The very thing we live in, experience every moment, depend on and are enslaved by, is something that remains a mystery to us.

It's... Shocking. Ridiculous. Insane.

And at the same time, it makes total sense. How can one see the glass jar they are in when glass is how they define their reality? Everything they know, everything they experience, everything they understand, is built only on what they are aware of. They simply don't know what they don't know.

Clearly, awareness isn't enough. Knowing there is an illusion isn't enough. It's why everyone who claims time to be an illusion cannot explain why. Seeing there is a veil of illusion doesn't mean you can see past it. And that's what I realised; the cloak of invisibility can only be lifted once you understand what's holding it down.

So that's where I started. Understanding.

Level 1 – the first level of understanding of 'time' – begins with diving into what time truly is, right from its most basic elements. Funny thing is, to do this, we have to go back in time, back to when the concept of 'time' was invented by humanity. Or rather, when our species became aware of it.

The Origin Story of Time

When you think of time, you probably think of clocks. So, let's start from there.

We all depend on clocks, watches, the 'time' to track and plan our days. We set our alarms to go off at a specific time, have breakfast, lunch and dinner at roughly the same times each day, aim to work for specific durations of time, and plan when to sleep to get enough time for a good night's rest.

In this context, what on earth is time?!

Let's look at it in its most basic form. What do we see on a clock? An example would be 7.30pm, or 19:30. Numbers that represent hours and minutes. But what is an 'hour'? What is a 'minute'?

This is where we go back in time, 5,000 years ago, to where clocks were invented.

Enter, the mighty Egyptians.

Have you ever heard of a Shadow Clock? Sounds cool, right? It was effectively version 1 of what became the Sundial. If that's unfamiliar to you also, it was a type of clock that used the shadow formed by sunlight to measure time.

And that's where the first revelation of time begins to shine through.

Imagine a stick placed into the ground, positioned upright. As the Sun moved through the sky, the shadow of the stick on the ground would move. Now, imagine 12 markers along the path the shadow takes. As the shadow moved to touch each marker, you could tell which hour it was. 12 hours of daylight.

THE THEORY OF EXTUITY

If you're wondering why they chose to split it into 12 parts instead of 10, I did too. It turns out that the Egyptian numbering system didn't use the decimal system we use today, based on '10s'. They used a duodecimal system, based on '12s', and there are two reasons why.

One is the fact that it equals the number of lunar cycles in a year, the number of times the Moon orbits the Earth. Astrology was very important to them. The second is due to the number of finger joints on a hand, excluding the thumb; that's three on each of your four fingers, 3x4=12, meaning the thumb can tap each finger joint to count to 12. Makes sense now, right? (Fun fact: duodecimal still exists today in other places, such as 12 inches per foot.)

So, our 24-hour clocks today originated from the Egyptian's Shadow Clocks and Sundials. Yes, clocks were invented based on the movement of shadows.

That means hours, or rather, time, has nothing to do with numbers. We can't even say it is 'related' to movement. In fact, it *is* movement. Now, don't get confused here. Remember it's not actually the Sun moving, it's the Earth spinning. Angular momentum.

Everyday time is a measure of the Earth's angular momentum.

LEVEL 1

Let's look at that a little closer. It's time for some physics talk. Momentum is proportional to the square root of kinetic energy. To have momentum is to have kinetic energy. Kinetic energy is the *energy of motion*.

So... time is the energy of motion?

Hmmm... let's pause that trail of thought for now, there's something else our history has to teach us.

Many believe that the origin story of time begins in ancient Egypt, the earliest known inventors of the 'clock', but it doesn't. That's because clocks weren't actually the first form of time measurement. To understand, we have to go further back in time. Much, much further back.

11,000 years ago.

Humanity was maturing. We grew from living as hunter-gatherers to living in settlements, introducing farming and domestication of animals. With this advancement came architecture, which takes us to the remarkable Tek Tek Mountains in Turkey.

Enter, the Karahan Tepe archaeological site.

THE THEORY OF EXTUITY

Throughout history, we've discovered many ancient sites that have been designed to align with the Sun on summer and winter solstices, the longest and shortest days of the year that mark the changing of seasons. Stonehenge in England, Machu Picchu in Peru, Chichen Itza in Mexico. And, it seems, Karahen Tepe may have been another, and the oldest of them all.

What does that have to do with time?

Once again, we see shadows cast by the Sun being used to track time, but in this case, on a much larger scale than daily tracking. Not hours or days, but years. Not clocks, but calendars. Specifically, solar calendars. The largest measurement unit of time humanity has invented for everyday life. It's what our modern calendars are based on.

Oh, and guess what, it seems that 1,000 years later another type of calendar was invented by mankind. A monument dated over 10,000 years old was discovered in Scotland that appears to track the phases of the Moon, the 12 lunar months in a year. Specifically, lunar calendars, inspiring the modern-day Chinese and Islamic calendars.

So, we have solar and lunar calendars.

Both types of calendars invented by mankind are measures of movement, but these aren't based on the spin of the Earth. No, they're based on orbits. The solar calendar is based on the Earth's orbit around the Sun, and the lunar calendar is based on the Moon's orbit around the Earth.

LEVEL 1

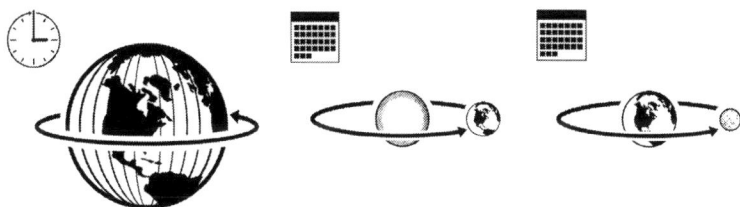

It's critical to understand that both measurements of time – clocks and calendars – are both based on completely different sources of seemingly-constant movement; Earth's Spin vs Orbits. They are not the same type of time. That's why leap years exist (every 4th calendar year adds a day to February to have a total of 366 days instead of 365 days), because a year is actually 365.2422 days (roughly 365 and ¼ days). Earth's spin, a 'day', has no connection to its orbit around the sun, a 'year'.

The 2 types of calendars – solar and lunar – are also different types of time. It's why solar and lunar calendars have months of different lengths; there are 12.4 lunar months in our normal calendar year, a solar calendar, because Earth's orbit around the Sun is completely unrelated to the Moon's orbits around Earth.

Yet surprisingly, while all of these types of time are different, they are all based on the exact same type of movement. The Earth's and Moon's orbits, just like the spin of the Earth, are once again angular momentum. Momentum. Kinetic energy. The energy of motion.

Yes, once again, we've reached the same result.

Time is the energy of motion.

The flow of time *is* the flow of energy...

THE THEORY OF EXTUITY

...which reveals a beautifully simple yet truly powerful definition of time...

Time *is* energy.

Time's Journey Through Time

You may not understand the implications of time literally being energy, and that's totally fine. I didn't at first either, until I dived into one of the most confusing and shocking discoveries of time…

Time Dilation.

Now, very few people understand what this is and, more importantly, how it works, but it's vital that you do in order to understand 'time'. So, I've drawn up diagrams to help simplify the explanation. Before we get into that though, let's start with a little story…

It's the year 1905, in the month of May.

A young patent clerk, on his way home from work, looks up at the famous Zytglogge clock tower in Bern, Switzerland. He enjoyed running various "thought experiments" in his mind that questioned and challenged accepted truths of the universe, appreciating the limitless power of imagination.

This evening, his creative mind manifested an unimaginable scene.

The scene depicted a tram car travelling away from the clock tower at the speed of light. Now, as we know, 'light' is how we see everything around us. Right now, light is reflecting off of these words that you are reading and travelling to your eyes at its natural speed, the "speed of light". So, the question he challenged here was: What if we travelled at the speed at which that light reached our eyes? Effectively, time would seem to 'freeze'.

This possibility blew his mind.

THE THEORY OF EXTUITY

This meant that for someone travelling closer to the speed of light, time on the clock tower would appear to run slower. Of course, this wouldn't make sense as everyone still at the clock tower would see the time ticking away at the normal pace. So what was going on here?

This led him on a time-bending journey revisiting the discoveries of Isaac Newton's "Laws of Motion" and James Clerk Maxwell's "Laws of Electromagnetism" defining the speed of light. His conclusion was shocking, an undeniable truth that changed not only how humanity saw 'time', but how we saw the universe.

He discovered that the closer an object travels to the speed of light, the slower its time flows in relation to a stationary object. In other words, time for that object slows down; the world around it would see it ageing in 'slow-motion'.

He discovered the Theory of Relativity.

His name? Albert Einstein.

To explain this a little better, it will be easier to compare two identical clocks in each scenario:

Imagine two identical clocks. One is on Earth, stationary, and the other is in a rocket travelling close to the speed of light. For the high-speed clock in the rocket, time feels like it is still flowing normally, but it's actually ticking slower in comparison to the stationary clock; time has been stretched and has slowed down. Likewise, for the stationary clock on Earth, time also feels normal, but in reality, it's ticking faster than the high-speed clock; time is squeezed in comparison and is faster to flow through.

As a result, a minute on the high-speed rocket is slower than a minute on Earth. In other words, the faster you go, the slower your time flows; the slower your energy flows.

What's quite interesting about this phenomenon is that if you were in the high-speed rocket, you wouldn't feel time acting any differently, it would feel completely normal. However, if you compared your age to someone on Earth, you'd discover that you would have aged slower than them. Time *around you* would have ticked faster in relation to yours.

This isn't fiction, it's proven science.

Time is elastic.

This is the critical discovery that the Theory of Relativity revealed to us about time; it can be stretched and squeezed. 'Dilate' means to widen, hence "Time Dilation". How? By accelerating an object, increasing its speed. In other words, by increasing its momentum.

Momentum. Kinetic energy. Energy of motion.

Once again, time is dictated by the energy of motion. The greater an object's momentum, the slower time flows in relation to an object with lower momentum.

However, it didn't end there.

In 1907, just 2 years later, Einstein described another phenomenon where Time Dilation occurred, but this time it wasn't caused by speed. This time, it was caused by gravity. As a consequence of his Theory of Relativity, he further concluded that the stronger the gravitational force on an object, the slower its time flows in relation to an object experiencing a weaker gravitational force.

To put this into perspective, this means that time ticks faster at the top of Mount Everest, nearly 9000 metres above sea level, where gravity is slightly weaker higher up in the atmosphere compared to everyone else on the ground closer to sea level.

Crazy, right?

Well, as it turns out, this was confirmed in 1959 by the "Pound–Rebka" experiment, but don't worry if you haven't heard of that.

LEVEL 1

Funnily enough, it has also been proven by a piece of technology that we use every single day.

GPS.

Yes, really. Our road navigation technology.

When our smartphones and car navigation systems utilise GPS, they connect to 3 of the nearest GPS satellites of the 30+ orbiting in Earth's upper atmosphere. Now, the mathematics used to calculate your accurate position on Earth requires very precise clocks in each satellite, which is where a very unexpected problem was found.

The GPS clocks could not be the same as our clocks.

If they were, our GPS locations would be increasingly inaccurate by over 10km every 24 hours. In a week, our phones would show us over 70km away from our actual location. Why? Because the clocks on the satellites *above* Earth would tick faster than those *on* Earth.

Einstein was right.

Just as he predicted, time down here in a stronger gravitational field ticks slower than time in the upper atmosphere where gravity is weaker.

So, because time speeds up in weaker gravity, GPS satellites have special clocks that are programmed to tick slower than normal in order to tick at the same speed as clocks on Earth.

Proof. Clear, undeniable proof that time is *not* a universal constant.

And that's part of the illusion.

But let's hold that thought for a second... there's something else that gravity is showing us here. We now know, without any doubt, that gravity can speed up and slow down time, "Gravitational Time

29

THE THEORY OF EXTUITY

Dilation" as it's known, but there's something different about it this time.

Gravity shows us that an object doesn't have to *move* to experience Time Dilation.

Wait... what?

Let's look at this a little closer. We know that time for an object will tick slower the closer it travels to the speed of light, but now we also know that time for a stationary object will tick slower the greater the gravitational force acting on it. Not kinetic energy, but gravitational potential energy. So, the fact that an object is moving doesn't matter for 'time', it's just about the force acting on it. It's just about the energy an object is experiencing.

So, time isn't dictated by energy 'of motion'. It's purely the energy acting on an object.

Time. Is. Energy.

Unchaining the Veil of Illusion

One of the biggest causes of time being a stubborn illusion is due to how we experience it day to day, but understanding 'time' that way is like trying to understand that we're living in a glass jar when glass is how we define our reality.

In order to lift a veil of illusion, we must understand what's holding it down, and the illusion of time is being anchored down by how we've come to define it.

Our clocks and calendars.

We live, day by day, with the belief that time flows at a constant rate. We see numbers on a screen incrementing in time with the spin of the Earth. We see nights come and go in time with weeks passing into history on our calendars.

We believe that if we haven't managed to do what we wished to in an 'hour', we've 'lost' time. We believe that each day we wake up, we have a certain amount of 'time' to work with. We are taught in life that we will have a limited amount of 'time' to live the lives we wish, to achieve the wins we hope for, to experience the wonders of the world around us.

But... it doesn't really work, does it?

We experience moments where time seems to fly by or drag on. Weeks where days felt like months, or seem to not have existed at all. Decades where nothing happens, and weeks where decades happen. Years that hold an entire lifetime of joy and success, or absolutely no life at all.

We live by universal numbers, instead of the individual, personal passage of time we have.

It's not just an illusion, it's a misdirection.

Time is not chronological, it's experiential. It's not universally constant, it's unique to each and every one of us. It's not numbers, it's energy.

And guess what, I'm not the first to suggest this. Humanity discovered this many moons ago. A people, and a language, have held a forgotten secret to time that has hinted at this very same truth.

The Ancient Greeks.

Very few people know that the Greeks actually had 2 different words for time. Kronos, and Kairos.

Hold on a sec... two different words... for 'time'?

Yes.

And what's funny is that no one knows how to explain it properly because of how we, today, see 'time'.

Let's look at this together...

'Kronos' time is chronological time. It's how we define time today; chronologically, continuously ascending numbers. Clocks and calendars. It ticks on and on. It can be quantified and measured. It's linear, moving from the infinite past into the infinite future. It has no freedom.

So, what on earth is 'Kairos' time?

Kairos is experiential. It's personal. It's not quantitative, it's qualitative. It's not a universal constant. It's not the same for everyone.

Kairos... is energetic. It's alive.

Kairos is the true meaning of time.

The illusion? Kronos. Calendars and clocks. But, it's an illusion we need. Kronos time is the glue that connects all of the differing Kairos times. Clocks and calendars are how we track the different flows of time that objects experience at completely different points in space, experiencing different levels of energy.

And now, it's time to begin setting the veil of illusion free.

Time is not a universal constant.

Time flows at different rates throughout the universe, because energy flows at different rates throughout the universe. Time is not constant, because energy is not constant. Time is unique to every object, because energy is unique to every object.

Time is not *like* energy, time *is* the resultant energy an object experiences.

The flow of time *is* the flow of energy.

Time is energy.

It can be stretched and squeezed. It can be sped up and slowed down. It's proven to be influenced by the speed and gravity acting on an object. That means if humans lived on a different planet with a different strength of gravity and orbiting at a different speed, they'd age at a different rate. If life really did exist out there in the universe, they could advance 10 times faster or 10 times slower than us.

And, a revelation that few consider... speed and gravity are just two forces that can speed up and slow down time... but, are they the only two? Could there be more "Dilation Energies"? And if there are, are there any that we, as humans, could control more easily?

Hmm... let's pin that question for now. At this moment in time, there's one critical learning to take away from this level...

Time is energy.

And guess what, this definition can be universally applied to *every* school of thought no matter how different their perception of time is. Physics. Psychology. Philosophy. Theology. Spirituality. It even explains how they have such differing views of time, because they are focused on different forms of energy. After all, there *are* different forms of energy, hence different flows of time, and the ultimate foundational definition for them all is, once again...

Time is energy.

This beautifully simple definition has powerfully vast implications. If you don't see them just yet, that's okay, this is just the beginning of our journey. This is a completely different viewpoint – or rather, a reality – to what your mind has been programmed to see since birth. It will all become clearer as you progress through the levels of understanding of time.

And to help with that, I'd like to introduce you to a new way of visualising the true nature of time.

As I've written a few times now, the flow of time is the flow of energy. For many generations now, we've referred to the flow of time as a "timeline", but I believe that is an incomplete representation of time. It was just a 'version 1' of what the flow of time really is.

Time is energy. A simple line doesn't capture the true nature of time. It should be something that represents energy, and what better than a wave?

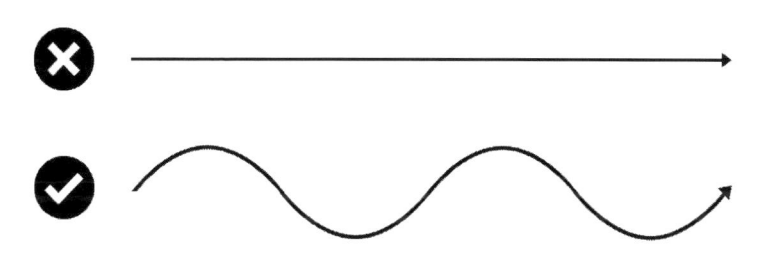

I believe it's the most perfect representation of time, not just scientifically, but psychologically and philosophically too. Life isn't straightforward. It isn't a straight line. It is, as we often describe it, a rollercoaster. You see it, right? Waves are perfect because waves come in all shapes and sizes, flowing at different rates, at varying frequencies and amplitudes. It's pure energy in all forms, scientifically, psychologically, and philosophically.

Not timelines. "Timewaves".

Timewaves perfectly capture the 3 key learning points of time that Level 1 has shown us:

1. Time is energy
2. Time is elastic; it can be stretched and squeezed
3. Time is not a universal constant

A timeline does not represent any of the above. A timeline does not represent time. Timewaves, however, perfectly represent each. Timewaves are elastic waves of energy that can be unique to each and every entity in the universe.

Not timelines. Timewaves.

THE THEORY OF EXTUITY

However, one incorrect assumption to make with a timewave is that it always flows at the same rate.

This is not the case.

While timewaves can be different for different objects in the universe depending on how slow or fast time is flowing for them, timewaves themselves can flow at different rates and intensities at different points in time.

In other words, the flow of time or the flow of energy of an object can change. A simple example of this is simply referring back to Time Dilation. Remember the comparison of a clock on Earth versus a clock in a high-speed rocket? Both clocks would have different timewaves; the one in the rocket would experience time at a slower rate in comparison to the stationary clock on Earth.

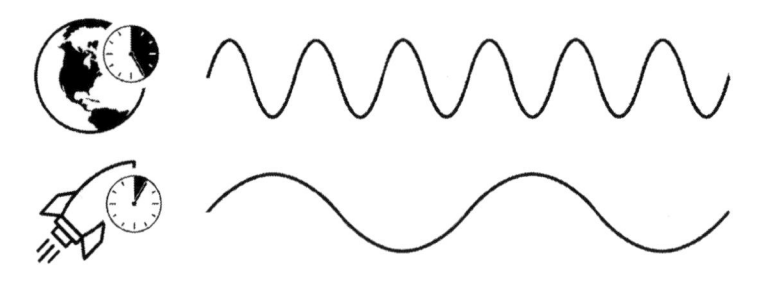

However, how would a timewave look for just one clock, a clock that starts stationary on Earth and then speeds off in a rocket? Time would initially be normal, and then would slow down in comparison to other clocks on Earth. This would result in a changing timewave.

This is fundamental to the nature of time. We know time is not a universal constant, but it is also not constant throughout its flow either. Why? Because time is energy, and energies can change.

And believe it or not, this applies to our everyday time too. Yes, I am indeed referring to the Earth's spin which dictates our daily time. The Earth does not and will not forever spin at exactly the same speed.

You've heard of leap years, right? Well, have you heard of leap seconds? They are one-second adjustments that are occasionally applied to our clocks due to the long-term slowdown in the Earth's spin. It was introduced in 1972 in order to account for the change in speed of the Earth's spin. In fact, since then, 27 leap seconds have been added to our clocks. That's nearly a half-minute adjustment to our clocks.

So much for a constant flow of time, hey?

Timewaves are a very important foundation of the understanding of time, but just a heads up, they are just the building blocks of the true, holistic, revolutionary representation of time.

Remember, this is just Level 1. This is just the first piece of the puzzle that will reveal an entirely new reality of time, or, quite simply, an entirely new reality.

So, hang in there, because we're just getting started.

LEVEL 1

Side-Level 1

Nature's Clocks

A very interesting question struck me when I was diving into the mystery of Time Dilation.

Let me take you through it...

For the clock on Earth, 'time' ticks at the standard rate. One could say at "Earth Time". For the clock travelling at high speed in the rocket, despite the clock programmed to tick at Earth Time, its flow through time is slowed down due to the speed it is travelling at. Time Dilation. So, in comparison to the clock on Earth, the clock in the rocket ticks slower.

Now, this I understand, but it is based on a very particular object.
It is based on objects with a "constant flow of energy".

So, what would 'time' mean for an object if its flow of energy was not constant? Objects that don't tick at the same rate? By the very definition of time in physics, those objects would have varying rates of time. As their clocks tick faster, their time speeds up. As their clocks tick slower, their time slows down.

Now, this is going to sound ridiculous, but bear with me. It will make sense.

Let's look at the same scenario but with different types of clocks, clocks that are programmed to tick at completely random speeds at completely random times. Now, the clock on Earth would tick at random speeds as randomly programmed, and the clock in the rocket would also tick at random speeds as randomly programmed, but overall slower due to the speed it is travelling at. Time Dilation slows all of its random speeds down.

As you can imagine, because both clocks were ticking at random speeds, we cannot compare the time on each clock. It's likely that the clock in the rocket would still have overall ticked slower, but we cannot know for sure. Both clocks would show completely random times.

Likewise, if these two clocks were on Earth without experiencing any Time Dilation, both clocks might show similar times, but they wouldn't be the same. Furthermore, the time on these clocks would be completely different to all other clocks that tick at normal rates.

But obviously, clocks like that don't exist.

Or do they?

Are there any such objects in the universe like that? Objects that have varying flows of energy?

Yes, definitely.

Every single biological entity on this planet.

Every human, every animal, every plant has varying flows of energy. So, by the very definition of 'time' in physics, every human, every animal, and every plant flows through time at varying rates. We sort of already know this, without knowing it, because we refer to it as our "Biological Clocks", but we haven't realised that it is in reality a different type of 'time'.

We haven't realised that "Human Time" has nothing at all to do with the time on our clocks, "Earth Time".

What's funny is that we have realised that 'time' differs between different animals, hence why we say "Cat Years" or "Dog Years". But again, Human Time and our Biological Clocks have nothing to do with Earth Time, and this points to one very interesting conclusion.

Our ages are a complete illusion.

In terms of understanding how 'old' we are, it's a very rough reference. Two people aged 30 aren't in fact the same 'age' because throughout their lives their energy flows would have been different. Being "30 years old" simply means you've been alive on this planet for 30 full orbits of the Earth around the Sun.

Again, your age in "Earth Time" does not accurately reflect your personal Biological Clock, your "Human Time". This means you are actually older or younger than someone who is the same 'age' as you. It's why two people dying of natural causes never die at the same 'age'.

Numerical ages are illusions.

Your personal energy flow is not constant. It cannot be compared to the energy flow of the Earth.

THE THEORY OF EXTUITY

Your time flow is unique to you.
So feel free to disconnect a little from your numerical age.
It truly means very little.

SIDE-LEVEL 1

YOUR JOURNEY
Level Progress & Key Learning Points

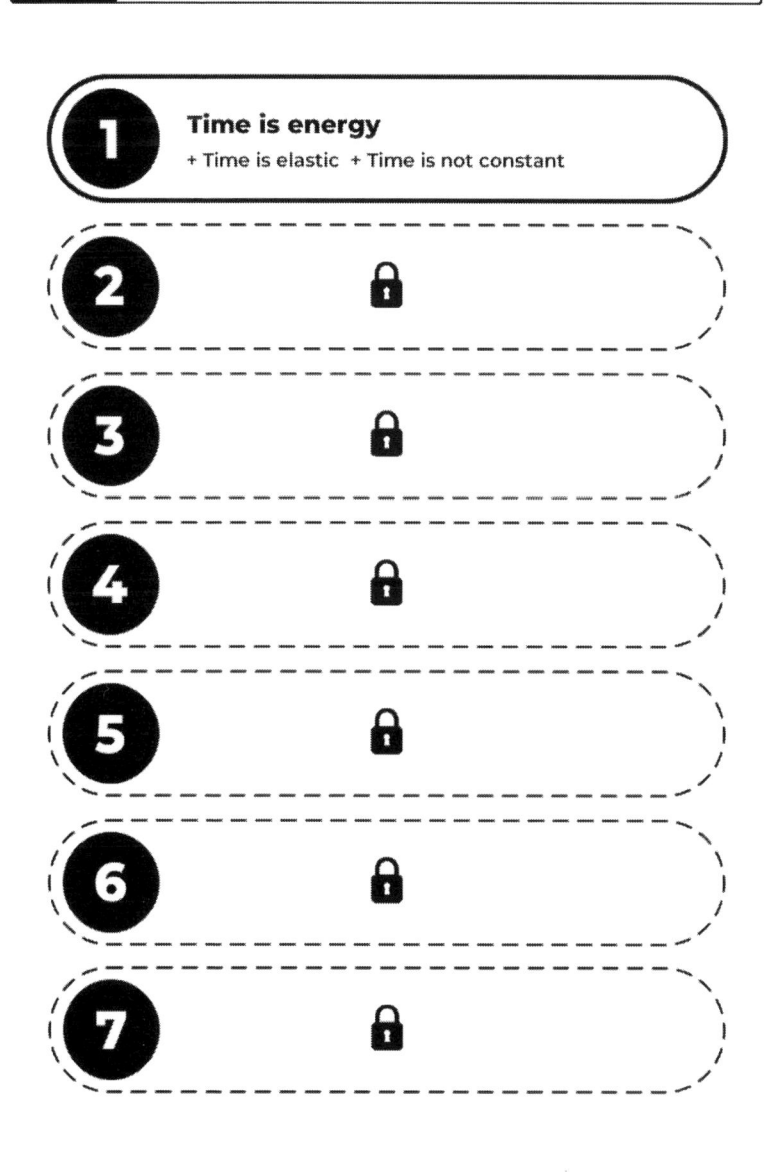

1 **Time is energy**
+ Time is elastic + Time is not constant

2 🔒

3 🔒

4 🔒

5 🔒

6 🔒

7 🔒

Level 2

Time travel is obviously... wait, what?

———

It's one of humanity's most captivating thoughts, not just within this topic, but in life.

Time travel.

You've probably seen movies around this concept. You've probably even dreamt of travelling back to a specific point in time yourself. So, the big question is: Is it possible?

As my friends and family would tell you, believing anything is possible lies at the core of who I am. It's unbelievably important in life. Why? Because if you don't believe something is possible, you'll never find out if it is. It requires every ounce of effort to prove a seemingly impossible reality is indeed a possible reality.

If you firmly believe something is impossible, even if a door of possibility existed, you'd never find it because you'd simply never look. You'd walk right past it. Your mind would be completely blind to it because you would have convinced yourself it doesn't exist.

Let's not let that happen. Let's see what possibilities are waiting to be discovered. Let's keep that open mind and look for the doors.

So, with our shifting understanding of time, let's dive into a big question...

THE THEORY OF EXTUITY

Is it possible for *anything* to travel from one point in time to another?

Science's "WTF" Moment

Our journey so far points to a strong conclusion: Time is energy. So, let's look at the greatest topic on energy in the world of science…

Quantum Physics.

Okay, okay, I know, Quantum blah-de-gook is complicated! Don't worry, I'm here to help simplify things. I've won awards for user-focused design, specifically for designing concepts in a way for anyone and everyone to understand with ease, and I'm going to do my absolute best to apply that skill to Quantum Physics.

So, let's start with a beautifully simple first step: Understanding 'Quantum' as a word, and it comes with a surprise…

Quantum is derived from the word 'Quanta', which means 'energy'.

Yes! Quantum Science is just the study of pure energy. And yes, you're starting to see the tie-in with time. Time is energy, the flow of time is the flow of energy, and Quantum Science is the study of energy.

Doesn't sound so scary now does it?

The reason Quantum Physics is known to be such a complicated topic is simply because the discoveries within the topic have been so mind-bendingly inexplicable. As a result, it has built a reputation of being something no one will understand. I'm here to disagree with that.

Did you know that we still don't understand 'Gravity'?

Seriously.

We understand its effect on our reality, sure, but we don't understand how it's created, why it has the effect it does, or what it

THE THEORY OF EXTUITY

truly is. That's why we still cannot generate, manipulate or play with it in any way we'd like. Yet no one says it's complicated, right?

And why is that?

Because we see it, feel it, live in it. Because we experience it. But guess what, it's the same with Quantum Physics, but the links with our daily experiences just aren't as clear, and I was blown away when I realised how relevant it was to our human experience.

It was a critical step to me discovering the Theory of Extuity.

And now, on Level 2 – the second level of understanding of 'time' – we are going to take the first, very critical step to understanding the connection between Quantum Physics and our human experience. And that, my friend, starts with one of the most mind-bending discoveries in science.

It's time to go back in time again, before Einstein was even born.

Just over 200 years ago.

The year 1801.

In 1801, there lived a scientist called Thomas Young. He performed an innocent experiment where he shone light through 2 parallel slits, the "Double-Slit Experiment".

Now, the wave nature of light means that it can act like ripples in a pond. So, as Young directed one light source towards the two slits, he expected the light passing through to interfere with each other, and it did. It produced an 'interference' pattern on the screen opposite, a simple striped pattern made up of bright and dark strips.

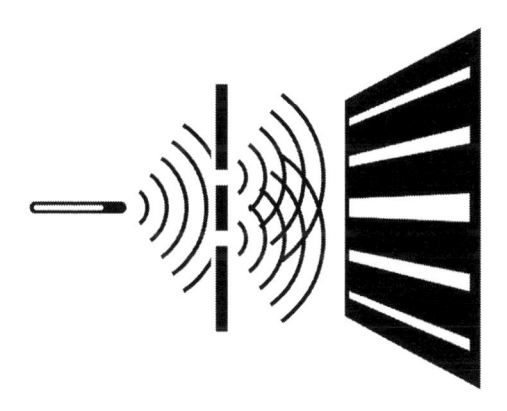

It was a simple experiment, one that you can easily reproduce at home, but it's one that opened the doors to a truly mind-blowing discovery.

Inspired by Thomas Young's double-slit experiment, a wave of scientists over the coming 2 centuries used the double-slit model to test many theories. Instead of using light waves, one of these experiments tested quantum particles – specifically 'electrons' – passing through the slits instead.

It shocked everyone.

As particles usually behave, similar to bullets from a gun, two lines were expected to appear on the screen opposite, representing the two streams of particles passing through the two slits.

THE THEORY OF EXTUITY

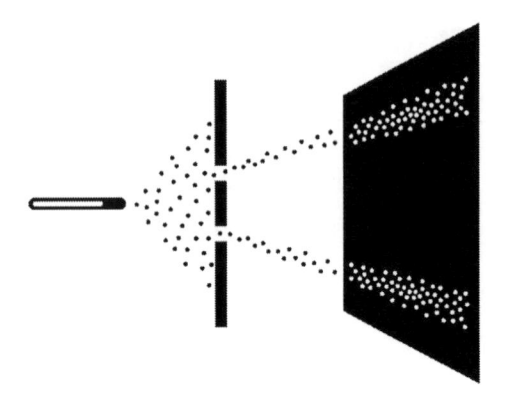

However, that's *not* what happened.

When particles were fired through the two slits, it was discovered that an interference pattern was created on the screen. Somehow, the particles were acting like waves, like ripples in a pond rather than bullets fired from a gun.

It made absolutely no sense.

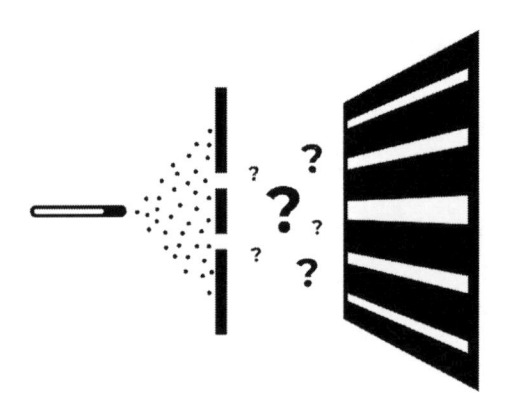

Of course, things got weirder.

To figure out what was going on, why the particles weren't behaving like particles, a detector was placed at the slits to observe how they were acting as they passed through. So, the experiment was carried out once again, and the results made no sense at all.

This time, the particles 'decided' to behave like particles. This time, they produced only two lines on the screen as they were originally expected to.

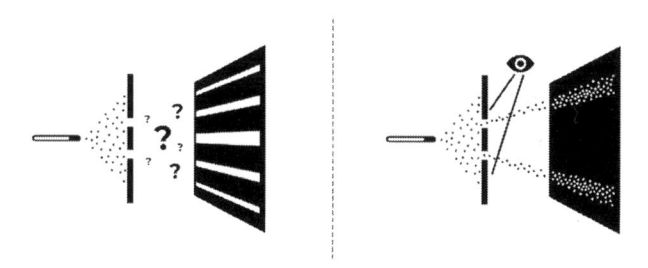

Confusion. Pure, utter confusion.

Why on earth were they acting like waves earlier? Maybe... maybe the presence of the detector was affecting the particles' behaviour? Well, it was possible. So the experiment was run again, and this time the detector was turned off but remained there.

So, the only difference now? The particles were simply not being *observed*.

And, guess what...

The particles acted like waves again (which they shouldn't be doing in the first place).

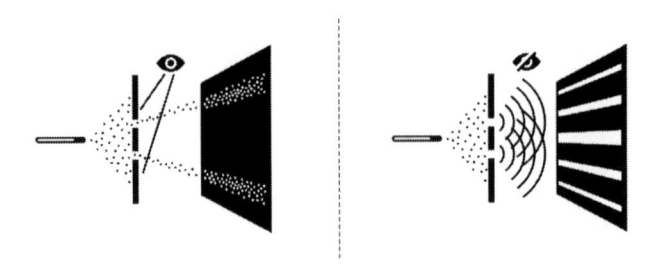

It made no sense. These are particles passing through the slits. They should be acting like bullets, producing two lines on the screen opposite. They should *not* be acting like ripples in a pond, they are *not* waves.

What did it mean??

Well, as crazy as it sounds, it turned out that the particles, when not being observed, passed through both slits at the same time and acted like waves. They existed as pure energy, pure waves.

In fact, they weren't *just* energy, they were in a state of being in *all* of their possible energy states. In other words, they existed in all possible realities, all possibilities of their existence, all possible states.

And when they were being observed passing through the slits? They only passed through one slit at a time, as we expect particles to act, as we see particles act in our reality. In other words, upon observation, they collapsed all of their possible realities into just one.

Or, to put into terms that describes our entire reality: Everything that you see before you only exists *because* you are looking at it, hearing it, feeling it, *observing* it in some way.

This isn't fiction. This isn't philosophy. It's an impossible-to-believe proven fact of reality.

Yes. "WTF" was the general reaction.

It's called "Quantum Superposition".

Think of 'superposition' meaning 'super-state'. For an entity to be in superposition is to be in all states that it can possibly exist in at the same time.

This has since been confirmed in many experiments of all kinds, including a recent Nobel Prize in 2022, and the conclusion is the same. It's called the "Observer Effect", or more technically, the "Wave Function Collapse", and it's one of the many experiments that proved Quantum Superposition.

Superposition is the ability to be in all possible states at the exact same time.

It's a fact of reality that humanity today still does not comprehend the implications of. It goes against so much of what we consider to be our reality, it's just too complicated to comprehend. However, as I always say, just because something is beyond human comprehension, does not make it impossible.

And that's where Erwin Schrödinger became famous for describing it in a way all would understand just a little better.

Enter, the Quantum Cat.

Yes, you did just read "Quantum Cat", and it's the most famous cat in science.

You've probably heard it by its other name, "Schrödinger's Cat".

Just to be clear, it's not a real cat, which you'll be very glad to know as you read on…

THE THEORY OF EXTUITY

The Nobel-winning scientist, Erwin Schrödinger, was a leader in Quantum Physics. Like all physicists, including Einstein, Schrödinger loved his "thought experiments", experiments that aren't feasible due to technical or moral limits, but have imaginative circumstances that still prove certain theories.

And that's where the cat made its debut, in Schrödinger's quantum imagination.

In fact, he came up with this thought experiment in 1935 in a discussion with Albert Einstein. His idea was to tie the fate of an animal with that of a particle in superposition. The best way to explain it is with the story itself.

Now, picture a cat in a box...

This is no ordinary box. This box is blacked out, completely opaque, perfectly soundproof. You cannot see inside it, you cannot hear any sound from it, you cannot 'observe' its contents in any way. However, you know a cat was placed inside, alive.

Yes, alive.

Ominous, I know. Read on.

This box, before the cat was placed inside, had a few objects placed inside it. A sealed flask of poison, a tiny radioactive source, a Geiger counter to detect the radioactivity, and a hammer.

Yes, things are getting weird, but let's continue...

Now, in the course of an hour, one of the atoms in the radioactive source could decay, but also, with equal probability, may not. It's in a state of superposition. However, if it does decay, the Geiger counter will detect it and release a hammer that will shatter the flask of poison, killing the cat.

Don't worry, there is no real cat, it's all imaginary.

However, in this scenario, Schrödinger successfully attached a real-world event with a quantum state in superposition. The radioactive particles may or may not emit an atom, but we have absolutely no idea because we cannot see or hear anything inside the box.

Because we cannot observe the inner state, the radioactive particles remain in superposition, in a state of having decayed and *not* decayed at the same time. Therefore, the cat is both dead *and* alive. In other words, the cat is in superposition, similar to a quantum particle in superposition; it is in all possible states at the exact same time until observed.

Only when the box is opened will one know if the cat is alive. In other words, only upon observation does the reality of the cat's fate collapse into one possibility. Just like the particles in the double-slit experiments, it exists in all possible states until observed.

The cat is, by definition, in superposition.

The cat is, by definition, both alive and dead until observed.

Or, as I came to realise, the cat is in two different states of energy, simultaneously.

Two different states of energy.

Yes, energy.

Do you get it?

Time is energy. Hence, it's in two different states of time.

Yes, exactly.

Schrödinger's cat is on two timelines – or rather, two timewaves – at the same time.

The Time Helix

Superposition, by definition, is the ability to be in all possible states at the same time. In other words, it's the ability to be in all possible energy states at the same time.

It's proven. There's nothing to deny there, but a startling realisation to accept.

It reveals an entirely new perspective of 'time' to one who understands that time *is* energy.

For an object, person, place, event, situation, or circumstance to be in a state of superposition is to be in all possible energy states, and by definition of 'time', to be on multiple timelines at the same time.

To clarify, this means that anything that we are not aware of, anything we have not seen, heard, or observed in any way, exists in all possibilities.

For example, someone may, at this very moment, be sending you a text message, or may not be. This means both realities exist until you see one play out. At that point, whether you receive a message or not, you'll collapse both timelines into the one that you've observed.

When I reached this point of understanding of time, a huge question arose in my mind, a question that humanity has been asking for as long as we've existed: Is it possible to influence the outcome? Is it really possible to "manifest" a reality? Or as others ask, is it possible to navigate our fate?

I couldn't find an answer right away, but when I stumbled across and finally understood the Theory of Extuity, the answer became quite clear.

The answer is both yes and no.

What I mean by that is that there is indeed a degree of influence over our reality, one that is extremely powerful, but not in the way people realise. There is a huge veil of illusion that masks the control we have. As a result, the majority of humanity fails to build the lives they wish, to 'manifest' in the realms of possibility, to influence their 'fate', all because of a misunderstanding of time.

The answer lies in truly understanding time.

What helped me greatly was adapting this concept of superposition to our improved understanding of timelines, which, as you know, are 'timewaves'.

Timewaves were just the foundation for something far more intricate and far more representative of our reality.

To introduce it, let's revisit our famous cat.

Schrödinger's cat, in the box, is in a state of superposition. It's both dead and alive until observed. It's in two energetic states at once. It's on two timelines, or rather, timewaves. One representing a reality with a dead cat, and one of a living cat:

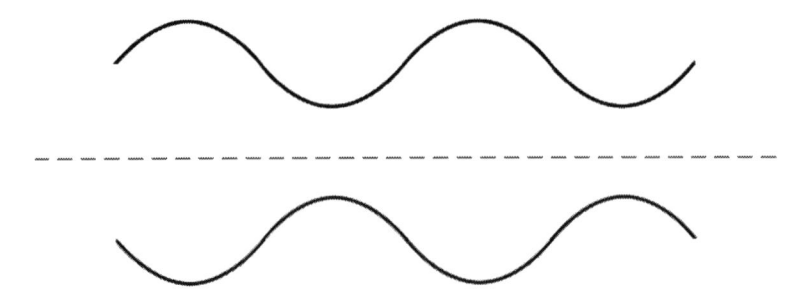

However, these two timewaves are not apart, they exist at the same time, in the same reality. This means that the actual representation of the cat's timewaves are of the timewaves overlapping each other:

However, there's one thing to remember here: the question of superposition here is focused specifically on whether the cat is dead or alive. It's been simplified to help explain superposition, but in reality, it's far more complex.

The true superimposed state of the cat would include far more than the two states of dead or alive. For example, what about injured? Sick? Tired? Energetic? Or what if we didn't even get to see the cat before it was placed in the box? What colour is it? How big is it? What gender is it? How old is it? There are, in reality, near-infinite states of possibility for this cat's existence.

Which means, as the cat could be in near-infinite energetic states, it isn't just on two timelines at once, it's on near-infinite possible timelines (until observed). Therefore, the true timewave representation of this infamous cat would include a multitude of timewaves overlapping each other:

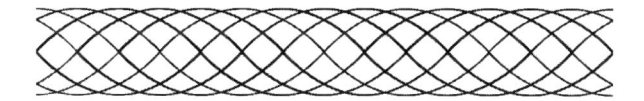

This, my friend, is no longer a timewave. This is what I call a "Time Helix". It's a true representation of our current reality, the "DNA" of time, and its implications are more powerful than you can imagine.

This model of time doesn't just apply to an unobserved cat. This applies to anything that we are not aware of, anything we have not seen, heard, or observed in any way, and also the limitless objects, people, places, events, situations, and circumstances that we have observed. There are timewaves for each and every aspect of our current reality.

The Time Helix is the most accurate representation of time.

Transcending Time

So, superposition gives time multiple dimensions, timelines connected *alongside* each other – side by side – but what about time connected in the other direction? Forwards and backwards?

Does superposition hold any revelations about time travel?

Most certainly.

Buckle up, because things are about to break the speed of light. Literally.

It's time we revisit the adventures of our dear Albert Einstein. Yes, you're probably starting to see why he was such a revered scientist, but there's a little surprise in this story.

This time, Einstein was wrong.

Yes, really.

However, don't think for a second that being wrong is a bad thing or something to be ashamed of. It was his stance on this topic that inspired reality-changing discoveries by scientists over the past century, discoveries that would have shocked Einstein himself.

To understand, let's head back in time to 1935...

30 years after Albert Einstein proposed the Theory of Relativity, his imagination was caught by superposition and he realised something shocking. He realised what Quantum Superposition suggested about how two 'synchronised' particles would act across space, and it was so shocking that he described it as "spooky action-at-a-distance".

What I mean by 'synchronised' particles are two or more particles that share absolutely identical energies. There are various ways two particles can be synchronised, with the most common being simply splitting one particle into two identical particles. Everything about each particle would then be connected, e.g. each would separate with exactly the same speed and momentum, and exactly the opposite spin and angular momentum. Synchronised energies.

However, following the laws of physics, the total energy would remain the same; the combined energy of the two child particles would be equal to the energy of the parent particle.

After all, energy cannot be created or destroyed; it can only be transformed from one form to another.

To help visualise this, think of a water droplet falling perfectly in the middle of a vertical knife's edge; it would be split into two identical smaller droplets, each half the size of the first, each falling at the same speed with the same momentum.

Their two energies would be identical, hence 'synchronised'.

Now, two synchronised particles aren't spooky at all, but when they are synchronised particles in *superposition*, that's when things get weird.

Let's apply this to the droplet example, but let's imagine it's a droplet of ink.

Now, if this droplet of ink is in a state of *superposition*, it means it is every single colour of ink at the same time; it's like how the cat was both dead *and* alive at the same time. So, when this droplet equally splits into two, both smaller droplets are also in superposition; they are both all colours at once.

What's "spooky" here as Einstein described it is that if you now look at one of the two droplets, it will collapse out of superposition and have only one colour; for example, it could materialise to be blue. However, the other droplet is *still* in superposition, meaning it is *still* every single colour at once.

Until you then look at this second droplet.

Once you observe the sister droplet, it will then have to decide on a colour, but according to the laws of physics, it *has* to be the same colour as the first droplet; it will somehow have to know that it *has* to be blue.

This is what Einstein realised.

He calculated that two particles in superposition that are *synchronised* means that if they were on different sides of the universe, and if they were both observed and forced out of superposition one after the other, they would somehow be able to 'talk' to each other instantaneously, *faster* than the speed of light.

By simply observing one, the other would *immediately* receive some form of energy and *know*.

Crazy, right? Einstein thought so. It simply didn't agree with the laws of the speed of light; *nothing* can travel faster than the speed of light.

In other words, it violated Einstein's Theory of Relativity. It violated the definition of 'time'.

So, Einstein proposed a different conclusion

He suggested that particles in superposition were given hidden information – or "hidden variables" as he called it – upon the point of being 'synchronised'. It was just an *illusion* that two particles could instantaneously share information across space faster than the speed of light, and in actuality they had that information all along, just somehow hidden.

He wasn't alone in thinking this.

Enter, Erwin Schrödinger.

Yes, the father of the Quantum Cat.

When he discovered this suggestion, he wrote a letter to Einstein in which he famously described these 'synchronised' particles as "entanglement", which this phenomenon has since been called.

"Quantum Entanglement".

So, two Nobel Prize winners… surely, they were right?

Surely, it's impossible for anything to *instantaneously* interact with each other across space? You can't affect the energy of a particle on the other side of the universe by interacting with a particle right in front of you, right?

In 1972, Stuart Freedman and John Clauser decided to find an answer once and for all.

What they discovered shocked the world of science, breaking all assumptions of reality, and proving both Einstein and Schrödinger wrong.

They managed to split one particle – specifically a 'photon', a particle of light – into two 'entangled' particles, and once a certain distance of separation was reached, the energy of one was measured and therefore forced into a specific energy state. The sister particle was then also measured to see if the total energy of both particles were the same, and they were.

No matter how far apart, the two particles were always able to instantaneously 'talk' to each other.

They proved there was no illusion to the phenomenon experienced between two entangled particles across space. They can, and do, somehow speak to each other instantaneously, faster than the speed of light.

They *proved* that energy can transcend the limits of space.

And that's not even the crazy part.

All of these scientists were so shocked at what it meant for our understanding of 'space', our physical reality, they missed what it meant for 'time', despite very, very clear signs.

A very critical aspect of these measurements went completely overlooked.

The entire focus was on what it meant for particles separated across space by a vast distance. No one looked at the timing of the observations. No one looked at the 'time' aspect of the experiment.

And that's where it gets really shocking.

In the experiments, it was assumed that upon measuring the energy of one particle, the energy of the sister particle would be affected. It was assumed that the first particle 'talks' to the second particle, telling it what energy state to take when it too was measured.

That's an assumption based on how humanity sees time.

If we throw a ball, we know we've written its future; its momentum will carry it in the direction we threw it and it will eventually make contact with something.

This is called 'causality', an event causing a future event.

Again, it's how we see time. Sequential. 1,2,3. Events happen in the order we see them, in sequence.

But, again, that's based on our current understanding of reality, a veil of the illusion that Einstein didn't realise was hiding a big secret in this mystery...

When we put aside our assumptions of time, we realise that by the very laws the experiment is based on, the first measured particle may not have been the one to 'talk' to the one measured later on. By the laws of Quantum Physics, when the second particle was measured later in time, altering its energy state, it could have caused the energy state of the first particle to change in the past, earlier in time.

In other words, there is absolutely no way to know at which point the energy states of the particles were affected; at the first measurement at the earlier point in time, or the second measurement at the later point in time.

Or, put simply, the future could have communicated with the past.

And guess what, this possibility was actually confirmed in a more advanced experiment of Quantum Entanglement.

In 2012, physicist Eli Megidish decided to take the experiment to another degree of complexity. Instead of analysing the energetic states of just one pair of particles, Megidish worked with two pairs of entangled particles in order to create an experiment that tested if energy states can be entangled across time with particles that didn't even exist at the same time.

In other words, irrefutable proof that entanglement across time is possible.

Or as it's now known, "Timeless Quantum Entanglement".

THE THEORY OF EXTUITY

And guess what... it was proven true.

Megidish concluded that there was no way to know whether the initial particle sent information to the final particle forwards in time, or whether the final particle sent information to the initial particle backwards in time.

There is no way to prove causality or retrocausality.

But, there is one fact that no one can deny; both points in time are linked in a way that completely breaks our understanding of time.

The same phenomenon that was discovered to exist across space was also discovered to exist across time.

Two different points in space can be connected, and two different points in time can be connected.

And that points to one, incontrovertible conclusion: All points in time can be connected.

In other words, energy, the fundamental building block of reality, transcends time.

Energy can time travel.

But... knowing it is possible is one thing, and being able to apply it is another entirely. When I originally read through Quantum Entanglement and understood everything you do now, I didn't make anything of it. I accepted it was very cool, but my logical mind concluded that it would only ever be a cool fact of science and not anything applicable to everyday life.

That was until I discovered a shocking similarity between the characteristics of Quantum Entanglement and the characteristics of an inexplicable phenomenon humans experience on a *daily* basis.

And now, it's time to take a detour on our journey through time. We've just completed our second level of understanding of time, and as I mentioned earlier, Level 3 is where we begin exploring what this all means for our human existence. Level 3 is where we begin diving into phenomena that we experience first-hand, revealing abilities we are unconscious of, and ultimately, learning how to master 'time' as a tool and a power.

It all revolves around a topic that humanity has long forgotten, stories that were once told by our ancestors, ancient teachings that modern society has lost.

Side-Level 2

Out-of-Body

I think everyone may have missed something in the double-slit experiment.

As we've covered, the "Observer Effect" shows us that upon observation, energy collapses into physical matter in our reality, becoming bound by space and time.

Is the opposite possible?

Is there such a thing as a "Non-Observer Effect", upon which a physical entity transcends into waves, pure energy, beyond the limitations of space and time?

When I thought about this, something came to mind…

There's a phenomenon that humans experience which many believe to be false, until they are one of the rare few to experience it.

An out-of-body experience.

You've most likely heard of it before. It's what commonly occurs during a near-death experience. If you research it, the common stories are of patients in surgery. When they wake up, they're able to describe the procedure to the stupefied surgeon, with their point of view of the events being from a top corner of the operating room.

Somehow, they 'saw' the procedure from a different point of view despite being unconscious.

THE THEORY OF EXTUITY

Naturally, common sense tells us it's impossible to 'see' if your eyes are closed, and most definitely if you're unconscious on an operating table. Yet, the evidence is undeniable.

It's a phenomenon that no one can deny.

In fact, the CIA utilised it themselves. Don't take my word for it, Google "The Gateway Program" and you'll find CIA documents released in 2003 of work conducted in 1983. These documents contain instructions on how one can meditate into a state where they can "see" beyond the limitations of their physical body at a different point in space; instructions on how one can intentionally induce the out-of-body state.

And what they describe has a commonality with all recorded out-of-body experiences.

Whether it's accidental, medically induced or attained through meditation, one has to reach a state in which they lose all awareness of their physical body and their immediate surroundings. In other words, reaching a "Non-Observer" state.

Could it be?

Could this be the opposite of the "Observer Effect"?

Upon total loss of observation of the physical reality, is an energetic state induced?

One to think about.

The Non-Observer Effect.

SIDE-LEVEL 2

YOUR JOURNEY
Level Progress & Key Learning Points

1 — **Time is energy**
+ Time is elastic + Time is not constant

2 — **All points in time can be connected**
+ One can be on multiple timelines

3 🔒

4 🔒

5 🔒

6 🔒

7 🔒

Level 3

Quantum ~~Physics~~ Biology

Level 3.

We're at a pivotal point.

We are now transitioning from mind-bending science to inexplicable human experiences.

Yes, the really good stuff.

This is the first of 2 key chapters. The first, Level 3, addresses a common human experience that everyone believes they understand, but few actually do. The second, Level 4, addresses a huge mystery of the human experience that not even the dictionary has an answer to.

And both levels address a timeline shift that defined the past 300 years of innovation.

When modern science discovered electricity in the 1700s, humanity's entire energy focus was pulled towards Physics, leaving Biology on the sidelines. Understandably, the application of electricity in technology held great potential, but our focus was redirected to understanding and harnessing the energy in the world around us and not within us.

It's a trick magicians are all too familiar with.

Misdirection.

Magicians are masters of illusion, and they understand and apply misdirection to achieve it with great effect. They trick their audience into being blind to something right in front of them by simply controlling their focus. But you have to remember, it's a skill that works because our minds are so vulnerable to it. Putting aside our ego, we are vulnerable to being blind to answers in front of our very eyes.

And in a world so much more connected, our collective eyes are prey to drastic misdirection.

So the big question is, what have we missed?

Ancient teachings of our human existence could have fallen to the sidelines, into history, long forgotten, and we'd be completely ignorant of it. Powerful truths could be right in front of our very eyes, or even more ironic, within our very bodies, and we'd be completely blind to it.

And Quantum Physics is an example of it.

Remember, 'Quantum' literally means 'energy'. Quantum Science has naturally been focused on Quantum Physics, but what about "Quantum Biology"? What about the study of energies that living organisms experience? What about the many inexplicable phenomena that humans experience?

Which brings us to the most misunderstood aspect of human existence... a key factor that few realise ties together *every* phenomenon that humans experience...

Emotion.

E-Motion

Did you know that the word "emotion" is derived from the Latin word *emovere*, a combination of "e-" (variant of "ex-") meaning 'out', and "movere" meaning 'move'? Essentially, it's derived from the phrase "move out".

However, today, 'emotion' has been given a different origin story.

The word "emotion", or "e-motion", is gaining traction in representing the phrase "energy-in-motion", and there's a very good reason why.

This time, I'm not going to start by explaining why. This time, we're going to dive into your human experiences where the answer has been all along. An ability that you have been unconscious of your entire life.

Are you ready?

Great, now...

Think back to a time when you experienced an *unexpected* emotionally intense moment. This can be a moment with a positive/high emotion, or a negative/low emotion. It doesn't matter where the emotion is on the scale, it just has to be a moment in time where you felt an *extremely* strong emotion, specifically one you didn't expect.

Now, let's teleport there for a minute.

Picture the place you were in. The room, the objects, the colours. Were there any distinct smells? Were there any specific sounds around you? Was anyone else there? There's no rush here. Close your eyes, take a moment to picture it all...

Got it? Amazing.

Now think about what was happening...

First, think about the event that triggered your strong emotion. Really put yourself back in that moment. And then, think about that emotion you felt. Feel it. Really feel it. Again, close your eyes if it helps, it's important you reconnect with the emotion in that moment.

Now, the big question...

Doing your best to put aside the thoughts that flooded your mind at that time, place your focus on how your physical body felt...

How did your body feel in that moment? What sensations did you feel? What changes in temperature, strength, or energy did you feel?

Does anything come to mind? If not, don't worry, as there's one more thing to try...

Really feel the emotion you felt back then. Name it, identify it, and let yourself feel it once again. But this time, feel it in *this* present moment right now. Really think about the event that triggered the emotion, and embrace it in this current moment...

Now...

How does your body feel? Can you feel an energy rise up within you? Can you feel an energy spreading to all corners of your body?

Yes? Amazing.

No? Maybe try this again, but pick a moment where you felt a really, *really* strong emotion.

Now, once you have felt the energy change within, it's time to dive into one of the most misunderstood aspects of emotions, and it can be simplified into one clear statement:

Emotions are labels for the *energetic states* of our bodies.

That's right.

'Emotions' are words we use to describe the energy firing through our nervous system.

Think about it.

When you experience fear, you feel a dropping sensation in your gut. When you experience love, you feel a warm sensation around your heart. When you experience nerves, you feel what is described as 'butterflies' in your stomach. When you're tired, you're low on 'energy'. When you're excited, you're 'energetic'.

Again, emotions are labels for our energetic states.

Yes, of course, there is the whole psychological aspect to it. The 'feeling' of fear is not just a dropping sensation in your gut, it's also how you mentally react to a certain situation or idea. The 'feeling' of love is not just a warm sensation in your heart, it's also how you mentally think of a certain person or place.

All emotions affect us psychologically, but each and every one has a physical reaction in our bodies. It's why we describe emotions as 'feelings', because they can be physically *felt*.

And we've been blind to its greatest secret.

The Ancient Chinese called it Chi. Ancient Indians, Prana. Ancient Greeks, Pneuma. Japanese, Ki. Tibetan, Lüng. Hawaiian, Ruah.

A powerful teaching lost in history.

There's a reason ancient cultures taught children to master their thoughts and emotions, and it wasn't for mental health. There's a reason ancient cultures taught children to meditate, and it wasn't for mental health. There's a reason ancient cultures taught children to identify, cultivate, and master their Chi, Prana, Pneuma, Ki, Lüng, Ruah, and it wasn't for mental health.

Mastering each was a foundation for our greatest human abilities.

Mastering each led to harnessing humanity's greatest power: Biological Energy.

Or, as we call it today, "Bioelectricity".

Our ancestors taught us to master our thoughts and emotions to more easily identify the sensations of our bioelectricity. We were taught to master meditation to cultivate and direct this biological energy. We were taught to master our bioelectricity so we could master powerful natural abilities, three in particular worth mentioning.

Three, which is said, takes monks years to master, and to the amazement of everyone around me, took me less than a year to understand, cultivate, and utilise.

The first ability?

Unprecedented power over our psychology.

We're all familiar with the fact that whenever we experience a certain emotion, we can feel specific sensations in our body. What we're specifically feeling is our bioelectricity spark and flow. So what happens when you experience it in reverse? What if you could consciously spark and control the flow of bioelectricity in your body? Believe it or not, you are able to directly recode your emotions.

Re-coding emotional responses is something that is indirectly achieved today through hypnotherapy or NLP (Neuro-Linguistic

Programming), but by having *direct* control of your bioelectricity, you have significantly more control over the ability of recoding emotions.

Those thoughts, events, objects, places, people that trigger emotions? Recoded. Reprogrammed. Gone.

The second? Well...

Have you ever wondered how low emotions affect our health? Why stress is especially harmful during pregnancy? Or what the 'placebo' effect is? Like I said earlier, emotions are labels for our energetic state. You might have already heard about it from ancient Indian and Chinese medicine, but in the Western world, we're only just starting to realise that it's our bioelectricity that directly affects our health, how strong we are against diseases, and how fast we heal.

Today we're taught to achieve these powers by managing our emotional health, but by attaining direct control over your bioelectricity, you have even greater control over your health. With *conscious* control of your bioelectricity, you can build your resilience to illnesses and speed up healing of all kinds. This includes directing bioelectricity around your body to reduce pain, headaches, motion sickness, and even speed up the healing of sore muscles, cuts, bruises, scrapes and more. It's everything our emotions unconsciously affect, but on a conscious level of control.

And the third?

It's the greatest power you can unlock by simply understanding bioelectricity...

Time.

What do I mean by that?

Remember, time is energy. And, by understanding that our emotions are directly related to our bioelectrical states, we realise that emotions represent our biological energetic state.

Or in simpler terms, emotions are energies.

Which reveals one of the greatest revelations of time.

Emotions are humanity's secret key to understanding our flow through time.

Bending Time

Humans don't like change.

We are fearful of anything that challenges our understanding of reality, but few know why.

You have, your entire life, had a certain view of life. You are, therefore, alive today because of that view of life. When that view is challenged, the natural response is to reject it. Why? Because, subconsciously, any new world is a world we have not experienced, and therefore a world we do not know we can survive.

It's a fear of death masked by our 'confidence' in the world we know.

And even if you are brave enough to embrace it, there's a second level of fear.

Accepting a new change to your world, accepting a new belief, would by definition separate you from your 'tribe'. This creates a subconscious fear of being kicked out of the tribe for being different as was the case for humans in the past, forced to fend for yourself in the wilderness. It's another fear of potential death.

When faced with great changes to the world we know, these ancient fears kick in.

When they do, there are two possible responses that you could take: Allow the subconscious fear to control you and reject any drastic changes to the world you know, or understand the fear to prevent its control and focus instead on the exciting possibilities of the new world.

In truth, we cannot master the world around us without mastering ourselves first.

It's important for you to keep all that in mind as we continue on this journey.

Now, it's time to dive into the first biological mind-bending aspect of time.

What's quite interesting is that phenomena in Physics are only discovered after diving into impossible questions, while in Biology it's the other way around. Inexplicable phenomena raise questions that we find impossible to answer.

I'm here to tell you that they are two sides of the same coin.

They are, in fact, mirrors of each other, and no one's realised it yet.

We're going to begin unveiling this connection with a very critical question that I asked myself when I began diving into time, when I set out on a mission to truly understand and remove the veil of illusion that blinded me to hidden truths.

So, I ask you now... When have you experienced 'time' acting unexpectedly?

Think about it.

It's an important question for any illusive topic. By highlighting these moments – specifically emotional reactions of surprise, shock, anger, frustration etc. – you'll highlight where your understanding of a topic is incomplete; there needs to be an adjustment, otherwise you'll stay stuck in the false reality and continue to experience moments that break your flow. More importantly, these moments can reveal critical clues to how something really works. So, I ask again, when have you experienced 'time' acting in a way that seemed odd, weird, or unexpected?

LEVEL 3

Does anything come to mind?

For me, a few came to mind, and the first was one we can all relate to.

Have you ever experienced any moments where 'time' seems to slow down, fly by, or freeze? Of course you have. We all do. And if you really think about it, and I mean, *really* think about all of these moments and compare them, you'll discover a connection between them all...

The flow of time 'seems' to change during moments of specific emotional states.

When we're having fun, time seems to fly by. When we're bored, time seems to go by extremely slowly. When we're in shock, time seems to freeze.

Time 'seems' to act differently.

There is of course an element of psychology here. When we're bored and constantly look at the clock, we're more aware of the flow of common time and hence more aware of its slow speed. When we're having fun, we're completely oblivious to the ticking clock and, in our ignorance of time, it flies by. And if you're familiar with the different brainwave states, that also has an influence here.

However, there's another factor in play.

Think about everything we've learnt. Think about how emotions are labels for our energetic states, our energy. Think about what 'time' really is and how it really works. Is there any proven element of time that shows how it can actually speed up and slow down?

Of course there is. Time Dilation.

89

As we covered in Level 1, the faster an object travels closer to the speed of light, the slower its flow of time in comparison to stationary objects, and as a result, time around it flies by. Moreover, the greater the gravitational force acting on an object, the slower its flow of time, and once again, time around it flies by.

This isn't fiction, it's a fact.

Science shows us that the greater the energy an object experiences, the slower its flow of time, and as a result time around it flies by. It works for speed. It works for gravity. But could it work for another type of energy?

Quite possibly.

The definition of Time Dilation doesn't just "kind of fit" our own experience of time slowing down and flying by, it fits to absolute perfection. Emotions, which are again just labels for our energetic states, could quite possibly be a third "Time Dilation" energy.

The identical characteristics cannot be ignored.

In Physics, the greater the energy an object experiences, time around it ticks faster and flies by. And as we experience, the greater our emotional energetic state, time around us ticks faster and flies by.

Or in simple terms...

The greater our emotional state, the faster we go through time.

Yes, seriously.

One could call it "Biological Time Dilation".

But don't let the phrase "the faster we go through time" fool you. Remember everything we've learnt about time. It doesn't mean a "universal time" speeds up because there is no such thing. Time is unique to every object in the universe. Every object has its own flow of 'time'. What happens is that *our* time slows down.

Effectively, our "Biological Clock" ticks slower.

Or, in other words, we age slower.

Think about it. You've seen this happen in reverse. When individuals go through periods of intense stress, they seem to age much quicker. In low energy, their biological clocks are ticking faster. It's what Time Dilation shows; the *less* energy an object experiences, the *faster* its flow of time.

And likewise, we see this happen the other way around too.

Countless studies show that a purpose-driven life significantly extends our lifetime. To have a purpose is to have a mission that electrifies every fibre in your body. To have a meaningful purpose is to experience a deep level of confidence, fulfilment and joy.

It cannot be denied that our emotional energies affect our biological time, *exactly* like Time Dilation.

It's just like the proven concepts of a clock in a rocket, and a clock in Earth's gravity. Except for us, as humans, the "Dilating Energy" here isn't caused by speed or gravity, it's the bioelectrical energy which we feel behind our emotions.

Our emotional energetic state seems to affect our biological clocks.

And if it does, can we control it? Can we consciously influence the speed of our ageing?

If so, there's a critical question to ask that is vital to understanding this a little better. Which emotional states are considered 'high energy' and which are considered 'low energy'?

THE THEORY OF EXTUITY

We have another great scientist to thank for answering that very question: Dr David R. Hawkins.

Released in 1995 after 20+ years of research, Dr Hawkins mapped a range of emotions to a range of values, a logarithmic scale from 1 to 1,000, which he titled the emotion's "Energetic Log". To be clear, these values aren't values for 'frequencies' or a specific measurement of 'energy'. These are comparative values to simply distinguish and understand which emotions are considered "high energy" and which "low energy":

ENERGETIC LOG			FREQUENCY
HIGH	700+	Enlightenment	HIGH
	600	Peace, Bliss	
	540	Joy, Serenity	
	500	Love	
	400	Understanding	
	350	Acceptance	
	310	Optimism	
	250	Neutrality, Trust	
	200	Courage	
	175	Pride	
	150	Anger, Hate	
	125	Desire, Craving	
	100	Fear, Anxiety	
	75	Grief, Regret	
	50	Apathy, Despair	
	30	Guilt	
LOW	20	Shame	LOW

Based on our everyday experiences, we know high *emotional states* make time around us seem to fly by. Likewise, based on Biological Time Dilation, the greater our *energy*, the slower our flow of time, and as a result, time around us flies by. And as you can see from this chart, emotions we'd consider "high" or "low" are indeed high and low energies.

Let's look at an example.

We all know time seems to fly by when we're having fun. Let's look at that emotion: Joy. Where is it on the energetic log? Near the top. It's a high-energy emotion. According to this, experiencing this would make time around us feel like it's moving faster than normal. Perfect, that's exactly how it feels.

Let's look at another example.

How about when we're really bored? The comparative emotion would either be 'Craving', as we crave joy or some stimulation, or 'Despair', and both of these are low-energy emotions. According to this, experiencing them would make time around us feel like it's moving slower than normal. And, as we all know, it does feel that way.

So, is it really possible?

Is it really possible that when we 'feel' time around us going faster or dragging by, it's because our physical being, even if to a very low degree, is going through time slower or faster? Does it make sense for nature to give us such a deep connection to our personal flow through time?

Yes.

It makes total sense.

Nature is ruthless. It makes sense for those living in higher emotional states, those closer to living in peace, joy and enlightenment, to have their physical ageing slowed, to live longer and help nature flourish. Inversely, it makes sense for nature to increase the speed of ageing of those living in emotional states closer to shame, guilt, despair; individually, humans and animals may not wish to prolong their lives in such a state, and collectively, nature would want to rid itself of potential dangers quicker.

Biological Time Dilation, with reason.

However, it's important to note that not all periods of time "flying by" are a result of Time Dilation, in particular when weeks, months and years seem to fly by. In these instances, time isn't being stretched or squeezed, time is actually in a 'loop'.

What I mean by that is, 5 years could actually be 1 year repeated 5 times. We could have lived each year doing the same thing. These are periods where our energy flow is in a loop, hence our time flow is in a loop. We do the same thing every day, every week, every month, and only when we break our schedule and do something different do we feel like we're living again, because, by the definition of time, we are. We break the time loop. We experience something new. We embrace a new flow of energy and truly add to our lifetime.

Hence the phrase, "It's not about the years in your life, it's about the life in your years".

But in those moments where time *really* feels like it has flown by or dragged on to the point it really shocks us, when we forget to eat for hours working on something powerfully fulfilling, when we're wired into a flow state and clocks mean absolutely nothing to us, our instincts may have been trying to give us a hint.

Time *really* was behaving differently.

Again, I'm not saying there is no psychological aspect to our experience of time flying by around us, I'm just pointing out the clear undeniable facts here. Science and nature point to the same *seemingly-impossible* possibility...

We, as humans, unconsciously speed up and slow down time.

The greater our emotional state, the slower *our* time flows, and the faster time flows *around* us.

Time is elastic, including ours.

Biological Time Dilation.

Energetic Bodies

I have a bit of a confession to make.

Biological Time Dilation, while being a really cool theory, has nothing to do with the Theory of Extuity.

So, why include it?

Well, like a lot of what we covered, it's a very critical understanding of time, but in this case, for one very particular reason. There is a very important teaching to take from it: Emotions, or rather, our energetic states, are intertwined with time.

When you really think about it, it shouldn't be that much of a surprise. We know all too well now that time is energy, and we also know that various forms of energy have the potential to bend and manipulate time. Speed, gravity, and, yes, bioelectricity, our bio-energy.

But... is it really *just* bioelectricity?

Let's revisit Quantum Physics for a moment.

Quantum Physics shows us that everything, absolutely everything in the universe, is made up of quantum energies. Everything in the universe has a pure 'energetic' version of it made up of subatomic particles and waves.

Everything, and I mean *everything*, has an "energetic body".

That's what Quantum Physics teaches us. But, what if we applied that very knowledge to Biology?

Quantum Biology.

Then, by definition, our very bodies have an "energetic body" counterpart to our physical bodies.

Or, in other words, a "Quantum Body".

Let's take a closer look at this.

If we were to zoom in on any physical matter, including our flesh, we would see all the quantum particles that make up the physical body. Electrons. Protons. Neutrons. However, we'd also find a *lot* of space in between the quantum particles. All physical matter, including our bodies, are actually 99.9% vacuum space.

But, it's not really empty.

Within that space flows the electromagnetic radiation emitted by our bodies, just like all matter in the universe emits, and also contains the electromagnetic field created by the electrical activity in our bodies. Both of these are waves. Both of these are pure energy.

Meaning 99.9% of any physical body is pure energy.

This is the energetic body.

And our energetic body is directly influenced by the electrical activity in our bodies. Our energetic body is affected by our bioelectrical energy. Our energetic body is connected to our emotions.

So, is it *just* bioelectricity that affects our flow through time? I believe not. I believe our energetic bodies also play a role *because* they are linked to our body's electrical activity. But there's something more important to recognise here...

We need to stop thinking that we are not related to the wonders of Quantum Physics. Our bodies are made up of physical matter and therefore subject to the same phenomena we've discovered in Quantum Physics.

We need to accept that we also have a "Quantum Body".

Once you do, the mysteries that we experience as humans begin to make sense. Answers to impossible questions are found. Realities we thought "impossible" are confirmed.

And when we begin to understand those impossible realities, we can begin to navigate them.

Which brings us to Level 4.

One of the biggest, most shocking truths of reality you'll ever be exposed to.

Or rather, one of the greatest mysteries of human existence that you'll finally understand to be one of our most powerful natural abilities.

One we have misunderstood for generations.

LEVEL 3

Side-Level 3

Our Quantum Bodies

I'm going to keep this really short because you're going to want to head straight to Level 4.

This is about Quantum Bodies.

We all have an energetic body, or a "Quantum Body". Each and every biological entity. Every single human.

Now, think about the abilities of any Quantum entity.

Quantum Physics proves that energies have abilities beyond time and space. Quantum energies don't necessarily experience time or exist in space as matter does in our physical reality.

Quantum Bodies by definition experience all those same mysteries. Quantum Bodies exist beyond the constraints of time and space. Quantum Bodies, if you think about it, sound exactly like something humans have described for as long as we've existed.

They sound exactly like how we describe 'Souls'.

Could it be?

Does Quantum Physics, once applied to Biology, scientifically prove the existence of souls? Is our "Quantum Body" our soul?

I'll let you think about this one.

YOUR JOURNEY
Level Progress & Key Learning Points

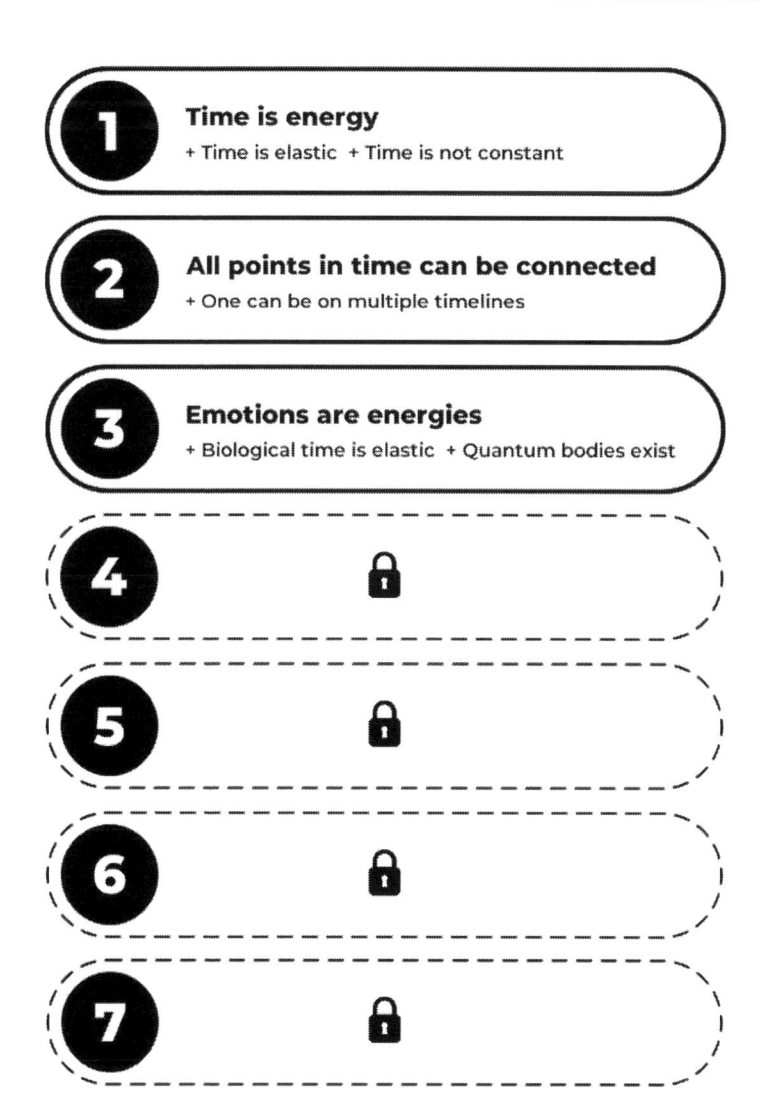

1. Time is energy
+ Time is elastic + Time is not constant

2. All points in time can be connected
+ One can be on multiple timelines

3. Emotions are energies
+ Biological time is elastic + Quantum bodies exist

4. 🔒

5. 🔒

6. 🔒

7. 🔒

Level 4

*The Word With A Bullsh*t Definition*

There's an accepted phenomenon of humanity that everyone experiences at least once in their lifetime.

And there are levels to its extremity.

There's one that everyone hears of, and then there's a second that not everyone does. The second, more extreme level is one that scares people, one that few share because of the fear of being ridiculed, belittled and shamed for how crazy it sounds.

But remember, crazy is what we want. Crazy is exactly what new discoveries start out as.

Crazy. Ridiculous. Incomprehensible.

So let's begin with the first aspect of this mysterious ability.

That Feeling

My older sister told me a story once, a story that I've never forgotten.

Back when she was a teacher, there was a young student – let's call her Sarah – in her class that was the absolute life of the group. Every day, without fail, she would radiate joy, cheek and excitement. There wasn't a moment, other than those when another upset her, that she was down or depressed in any way.

Except for a day that changed Sarah's life.

One morning, my sister attended school as usual. She had prepared her teaching plans for each lesson that day, ready to enlighten each and every student, and gladly welcomed her pupils to sit down for morning registration.

As always, Sarah was as bright and joyful as ever,

The morning began beautifully, the class was engaged and enjoying the lessons, and Sarah was her usual bright self.

Until midday came.

At some point around 12pm, Sarah's demeanour changed. Completely out of the blue, she was quiet and withdrawn. So much so that my sister noticed.

An hour or so later, the headmistress came to the classroom, knocked on the door, and asked for a moment with the teacher.

My sister felt something was off.

She let the class know she'd be back in a minute and went outside the classroom to speak with the headmistress.

My sister says she'll never, ever, forget that moment.

The headmistress let her know that earlier that day, around noon, Sarah's mother had died.

LEVEL 4

Through the shock that my sister experienced, something very clear clicked in her mind.

Sarah, somehow, felt her mother passing away. Sarah's entire state had shifted so significantly around the same time her mother passed, somehow, in some way, she knew. There was no logical way for her to have known, yet she did.

Or, to be specific, she subconsciously 'felt' it.

And if you've experienced the loss of someone extremely close to you, you may be able to relate perfectly well to what Sarah experienced.

It's not a fairytale. It's not fiction. Humanity is all too familiar with this. It's said that when we lose a loved one, we can feel it, no matter how far away we are from them. Yet, there is no logical explanation for it, so science dismisses it.

But it doesn't end there. It's not a phenomenon that's just tied to death.

Another common experience is when a loved one is in an accident. People just 'feel' something happened. They 'feel' a loved one is in great pain, fear, or shock. People somehow 'know' something has happened at a different point in space nowhere near their location. It's baffled scientists for as long as 'science' has been a topic.

In fact, despite it being something that has remained unexplained, it is such a deep part of our reality, the dictionary even has a word for it.

But, this is where things get really funny.

Because it has remained a mystery, it's a word that is technically defined as having no definition.

The word?

Intuition.

And before you think you know what it means, I urge you to remain neutral in how you see the word until you reach the end of this level, Level 4.

Intuition is a word that I looked up in the dictionary when I discovered how this phenomenon actually works, and it was then that I experienced my first moment of shock at how little humanity truly understood of our *natural* abilities...

Let's have a look at a few of those definitions.

The Oxford English Dictionary, the principal historical dictionary of the English language, defines 'Intuition' as "the immediate apprehension of an object by the mind without the intervention of any reasoning process."

The Cambridge Dictionary defines it as "an ability to understand or know something immediately based on your feelings rather than facts."

The Collins Dictionary, "unexplained feelings you have that something is true even when you have no evidence or proof of it."

And that's all they say.

There's no explanation for it.

If anything, they define it by stating what it isn't – "without the intervention of any reasoning process", "rather than facts", "unexplained", "no evidence or proof" – which, to me, is unbelievably ridiculous. *Where* does this 'knowing' come from?!

These definitions are empty.

They explain nothing.

They define no answer.

They are *undefined* definitions.

But... I don't blame them. The phenomenon is so common, so deeply accepted by society, it *had* to be included in the English dictionary. The problem lies in it being a phenomenon, a mystery, an inexplicable part of reality.

And, as I discovered, it has remained a mystery because we've yet to understand another mystery of our reality. An illusion that humanity has not yet understood.

The one that you are beginning to understand.

Time.

WTF.

Yes, WTF.

It's the reaction that hits people when they experience the second aspect of this mysterious ability, the second side to intuition.

And yes, there is a second element to it.

It's one that directly involves 'time', the very reason that makes people feel uncomfortable and scared to share their experiences of, simply because it breaks all understanding of time.

And it's one that you will be able to relate to upon hearing two simple words:

Gut feeling.

You've had a gut feeling about something before, right? There was no logical thinking behind it, no explanations for it, you just felt something had happened, was happening, or may happen, either to you, a family member, or a friend, regarding a certain event that was happening, or an event you had no idea about.

This time, I'll share one of my own experiences.

In fact, on the day I'm writing this, 11th of August, 2024, I experienced this very thing just yesterday.

I woke up at 9am, in Wales, in my brother's beautiful Welsh manor. I decided to head out here for a few days to spend time writing this book, surrounded by one of the most beautiful landscapes on Earth; green rolling hills bathed in golden sunshine.

However, yesterday was a special day. We were heading out to North Wales to climb the Cader Idris mountain, an 11km trek with a 900m ascent. It was going to be epic.

I jumped in my car, set off, and drove north along the winding country roads.

And just 5 minutes in, I had an unexpected gut feeling, a gentle tug on my mind. I looked further ahead down the road and felt uneasy about the junction I saw. It wasn't the first junction I'd be passing, but I had a feeling that a car would just pull out of this one without any awareness of me driving up the main road.

No logic. No reasoning. Just a feeling.

So, having built up a deep trust with my own intuition, I slowed down. I was in a 60mph zone, but I slowed down to 40, ready to brake and slow down further if I had to.

And, as I'm sure you guessed, a vehicle *did* pull out. An old pickup truck drove out from the sideroad ahead and drove straight across the road I was driving on.

I braked.

Thankfully, an emergency brake wasn't required because I had already slowed down earlier.

But... if I hadn't, it could have been a lot worse.

Seriously.

Yet, there was absolutely no logical explanation for my 'knowing'. It's exactly as dictionary definitions state, "based on your feelings rather than facts". According to everything we're taught as fact about our reality, about 'time', I should not have had any inclination of this future event.

It's crazy, ridiculous, *incomprehensible*.

And what I love about this is that when I started sharing stories like this that I'd experienced, others started sharing their own too.

And, oh my, were they absolutely crazy, in a wonderfully beautiful way. By sharing my own, I'd let them know that they were in a safe space to share their own experiences that modern society would label as crazy, when in reality many have experienced the very same inexplicable phenomenon!

One notable story was of a friend who, while walking up to a supermarket with her baby daughter, unexpectedly felt the urge to ask her partner to take the baby and wait. She'd *never* done that before. Later, as she left the store alone, a dog violently attacked her. She was left shaken, more so because she was picturing what would have happened if she was carrying her baby. Another story was of a friend who, as a child, suddenly had a panic attack in the car and begged his mother to stop; as they pulled over, a van with no headlights on at night went speeding around the blind corner just ahead. His mother was left speechless; if they continued driving, they would have been in a fatal accident.

If you haven't personally felt any intuition like this and haven't heard any stories like these, just ask people around you. Hopefully they feel safe enough to share. Many of us experience it on a daily basis, a "gut feeling" something is going to happen, was happening, or had already happened. A 'feeling' that could not be explained logically or with any evidence whatsoever.

And now, I'd like to point out a very specific aspect of these inexplicable stories of intuition, and to the definitions that dictionaries gave intuition.

There is a very specific word that is used in relation to each and every inexplicable moment.

The word?

"Feeling."

It's a word that relates to a very specific human ability. A word we covered in great detail earlier.

Emotions.

Yes.

Emotions lie at the heart of intuition. Sarah, the young student in my sister's class, experienced a very clear change in her emotional state around the same time her mother had died. When I was driving through Wales, I experienced a worry of a potential event that would occur, a fear of being in a dangerous situation.

As we've already covered, emotions aren't just psychological, we 'feel' them in our physical bodies. They are labels for our energetic states, driven by the bioelectricity coursing through our bodies.

Emotions are energies.

Emotional energies are what we feel when we experience 'intuition'.

It's a reality-changing fact that reveals one of the greatest revelations of our lifetime, and possibly even the history of humanity.

And to discover this revelation, there's one very specific Quantum phenomenon that reveals the secret…

Timeless

When I discovered what I'm about to share, I don't believe I've ever been so surprised, and at the same time, so sceptical of a truth that had slapped me so hard in the face.

I can still feel the emotion reverberating through time from that moment.

So, now, I would like to direct your attention to the two stories that I shared with you. Specifically, to the breakdown of 'intuition' that I've presented you with.

Two very distinct angles of the phenomenon.

One, with Sarah feeling emotions around the death of her mother, an event which she had absolutely no logical awareness of.

Two, where I myself felt emotions of the danger that was about to be caused by another driver, an event which I had absolutely no logical awareness of.

Do you see the difference in the two types of intuition?

One, Sarah felt an emotion regarding an event that happened at the same point in time but at a completely different point in space.

Two, I felt an emotion regarding an event that would happen at the same point in space but at a completely different point in time.

Do you see it... ?

One, energies are connected across space.

Two, energies are connected across time.

Remind you of anything?

In 1972, Quantum Entanglement across space was proven.

In 2012, Quantum Entanglement across time was proven.

We covered it in Level 2.

But, in this case...

One, emotional energies were connected across space.

Two, emotional energies were connected across time.

Somehow, "Biological Quantum Entanglement" is in play here. Take a moment to think about this.

"... emotional energies were connected across time."

Is it possible...?

Is it possible that emotions can bleed through time?

Is it possible that our emotions transcend time?

If it is... the implications are... unimaginable...

If you are beginning to see how radical a discovery this is, I assure you, the real implications of it are far greater than you can imagine. In fact, it's such a drastic shift to our understanding of reality, it took me months to process this, and much longer to understand what it really meant for humanity.

But, surely, if this is real, there must be some research, some evidence, to prove it?

I'm such a logical person, I had to find out. I searched endlessly, and in the end, I found it.

I found a scientific study that *proved* intuition across time.

Enter, the HeartMath Institute.

In 2004, HeartMath set out to validate the shocking results of a study conducted by Dr Dean Radin, senior scientist at the Institute of Noetic Sciences. It was a test to see whether participants could feel an emotional response to an event before it happened. HeartMath researchers refined Radin's experiment by measuring participants' brain waves (EEG), their heart's electrical activity (ECG), and their heart rate variability (HRV).

How did it work?

Participants were told they were in a study testing responses to emotionally stimulating photos, unaware of the true purpose. They pressed a button to begin each trial, after which the screen stayed blank for 6 seconds. A randomly selected photo – triggering either a strong emotional reaction or a calm state – was then shown for 3 seconds, followed by a blank screen for a 10 second "cool-down" period. When ready, they could again click the mouse to begin the next trial.

What they found was remarkable.

The results showed that both the participants' heart and brain received and responded to information about the emotional quality of the pictures *before* they saw them. Specifically, these responses occurred an average of 4.8 seconds before the computer randomly selected pictures, just 1.2 seconds after choosing to initiate the process.

Their bodies felt the emotional energy a picture would evoke *before* they even saw it.

What I found even more interesting is the fact that the data shows it was specifically the participants' hearts energetically reacting before their brains. HeartMath reported that the "intuitive signal" originated in the heart, was passed to the brain's emotional and prefrontal cortex, and then passed down to the gut, triggering "gut feelings".

But, we have to remember, or rather, I had to remind myself, that the only reason I wanted to hear factual evidence of this phenomenon was because of my ego. That deep survival trait, that subconscious fear of a world different to what I thought I knew and found comfort in, needed reassurance.

But reassuring my ego didn't change the reality of the situation.

In reality, I knew deep down that this discovery was right. I knew that I'd found a link between a phenomenon in Quantum Physics and Human Biology that would lead to profound implications.

Now, it was just a matter of acceptance in order to begin considering what it meant.

So, I kept on repeating the same line, over and over.

"Emotions bleed through time."

"Emotions bleed through time."

"Emotions bleed through time."

It would have so many implications for our understanding of psychology. It would change so much of how humanity understands "personality". It's...

... wait...

... hold on a sec...

Emotions... bleed through time?

We receive emotions from the future?

We can 'feel' emotions from events that haven't even happened yet?

So... our futures... could influence our past?

But... there's something else here...

There's something more profound that I'm missing here...

THE THEORY OF EXTUITY

Not long after all this hit me, my mind wouldn't sit still. It wouldn't allow me to sit and contemplate what I'd just discovered. There was something that it was trying to highlight, something it knew I was meant to question, something even more profound...

And *that* is when I discovered the Theory of Extuity.

LEVEL 4

Side-Level 4

Mind, Body, Soul?

———

As you probably know by now, I can be quite ambitious. So, you shouldn't be too surprised to hear that 'time' wasn't the only mystery that I attempted to find an ultimate answer to. There was another.

A blueprint for the structure of our "Mind, Body, Soul".

Why? Because the better we can understand the architecture of our very being, the better we can experience and make the most of the present moment, the here and now, with greater awareness, understanding and control of our existence.

For example, few realise how dependent we are on the Earth's electromagnetic frequency. In fact, NASA only discovered this when the first astronauts were sent into space; they experienced various health issues, including fatigue, stress, and compromised immune function, which they discovered to be a result of the lack of Earth's electromagnetic frequency. Upon creating a device that mimicked this natural frequency, astronauts no longer experienced these health issues when in space.

As we've covered, the Earth isn't the only object that emits electromagnetic energies. Our bodies do as well. This explains how some people can 'feel' the emotions of others without any awareness of their situation, often described as "Empaths"; their highly sensitive nature absorbs electromagnetic frequencies of others.

Don't get this confused with being empathetic. To be 'empathetic' is to understand how another is feeling, but to be 'empathic' is to feel the emotional energies another is feeling as if they are your own.

As we covered earlier, emotions dictate every single aspect of our lives, and emotions are a result of our bioelectrical energy. So, to truly master who we are, we must understand where and how energies flow between our mind, body and soul/quantum body.

We need an energetic blueprint. So, I attempted to design one:

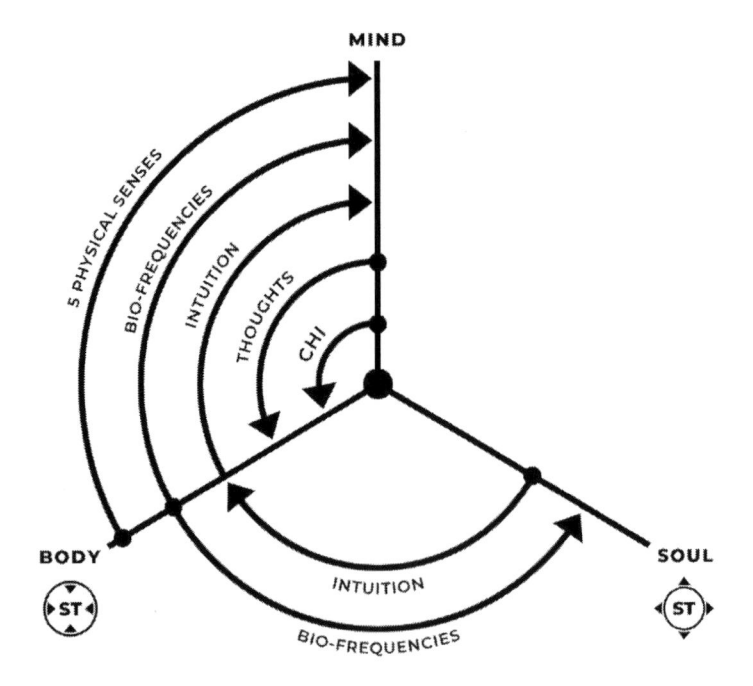

Let me take you through it.

First, I'd like to explain the "ST" symbols below "body" and "soul". This highlights the body being within "space and time", and the soul/quantum body being outside of "space and time", representing energy from both sources. I believe "mind" is unrelated to either, for example "imagination" is beyond space and time, hence it's a third source of energy entirely.

Now, the "5 physical senses" are the root of the majority of our emotional responses; i.e. if we see someone holding a knife, our mind will perceive danger and trigger fear. So, it flows from body to mind.

The second are the "bio-frequencies" we covered earlier; the electromagnetic energies we absorb, i.e. from Earth or people nearby.

Then there's "intuition". As we've outlined, it is connected to a source beyond our current point in space and time, and therefore must originate from our "quantum body" or "soul".

The fourth: Thoughts. For example, imagination or ideas can influence our emotional states.

Finally, the fifth, as Ancient Chinese Medicine calls it, is "Chi', our bioelectricity. Specifically, conscious control of it, the ability monks master after years of training.

However, there's a problem with this blueprint.

I believe its incomplete.

Why? Because there is absolutely no connection between the "mind" and "soul". It 'feels' wrong. My intuition is telling me something is missing.

However, I also feel like I'm not the one who's meant to complete this. I have a feeling someone out there has the answers that will fill in the missing links. So, for the first time, I'm going to leave this with you.

Think about it, share it, discuss it.

It could transform our understanding of humanity.

YOUR JOURNEY
Level Progress & Key Learning Points

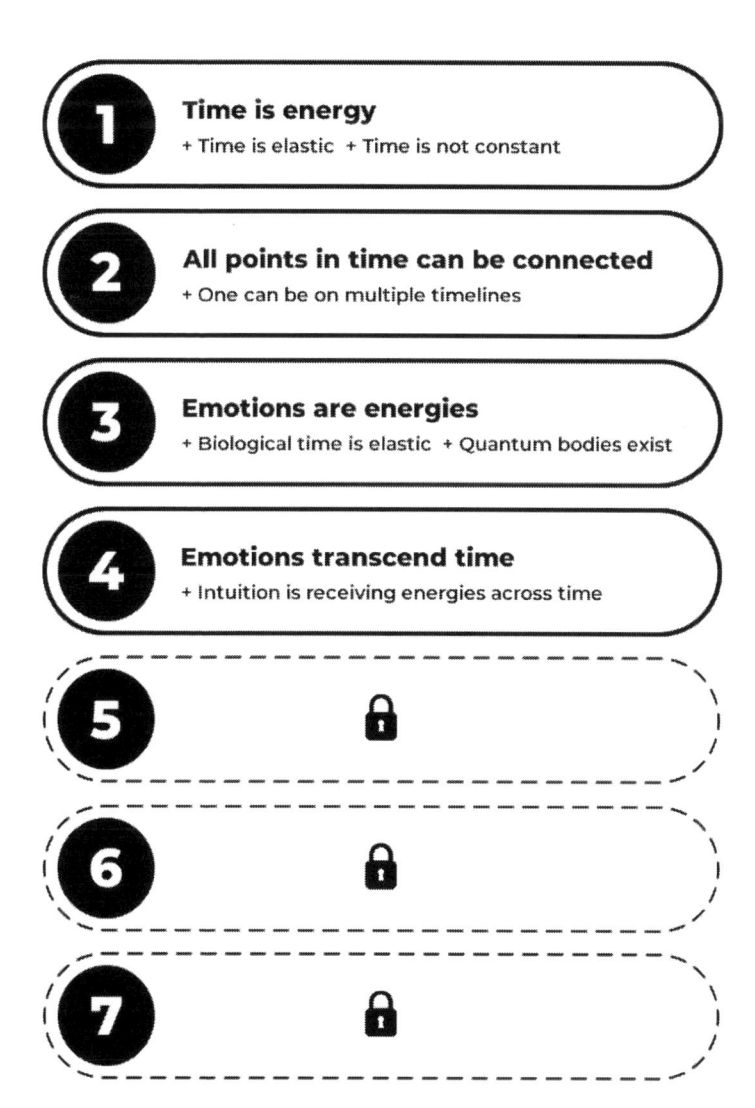

1. Time is energy
+ Time is elastic + Time is not constant

2. All points in time can be connected
+ One can be on multiple timelines

3. Emotions are energies
+ Biological time is elastic + Quantum bodies exist

4. Emotions transcend time
+ Intuition is receiving energies across time

5. 🔒

6. 🔒

7. 🔒

Level 5

The Theory of Extuity

Yes, finally, we arrive at the main event.

The Theory of Extuity.

"Extuity".

A word that *should* exist in the dictionary.

A word that I endlessly searched for in the hope of understanding it better...

... but, as it turns out, a word that was waiting to be found...

The Word.

"Emotions bleed through time…"

Every time I said it, every time I said the phrase, I felt like I was missing something really obvious.

"Emotions bleed through time…"

My mind was nudging me, pushing me, throwing me back at the revelation I had just come across, telling me there was something I was missing…

Intuition, Biological Quantum Entanglement, proved that emotions bleed through time.

The HeartMath Institute research proved it.

Intuition was, by true definition, receiving emotions from our future selves.

Wait…

I'm so close…

It's right there…

… we receive emotions from our future selves…

Yes, of course…

But then…

If we are receiving emotional energies from our future…

That means…

We are, by definition, right now, *sending* emotional energies to our past selves.

Wait…

Could that really be possible?

I need to circle back…

LEVEL 5

If intuition is the ability to receive emotional energies, what we 'feel', exactly as HeartMath Institute detected...

Then, by definition... what's 'received' has to have been 'sent'.

What's 'received' has to have been 'emitted'.

Obviously.

So... if intuition is *receiving* energies, what's the word for *sending* energies?

What's the opposite of 'intuition'?

Hmmm...

Internal, external.

Intrinsic, extrinsic.

Interior, exterior.

So...

Intuition... Extuition?

Yes...

That sounds right. The opposite of intuition is 'extuition'.

Intuition is receiving emotional energies across time, extuition is sending emotional energies across time.

But let's look up the actual definition.

Google, dictionary...

Search...

E...

Ex...

Scrolling... scrolling...

...

Nothing.

No 'extuition'.

Let's try another dictionary...

... and another...

... and another.

Nothing. No definition, anywhere.

WTF.

But... it makes sense...

I must be wrong.

Something in my logic must've been wrong.

Let me backtrack.

And I did... for months.

I went back and forth. I questioned everything. I tried to find scenarios, evidence, any reported experiences that disproved any part of the journey of understanding of time that I'd gone through.

But I couldn't.

I kept arriving at the same conclusion.

We, as human beings, experience the same phenomena discovered in Quantum Physics.

We, as human beings, experience Biological Quantum Entanglement.

We, as biological entities with "Quantum Bodies" or "Energetic Bodies", experience "Intuition", receiving energies from our future selves, and we've been gifted with the ability of translating and understanding those energies with our ability to feel emotions.

Leading to a theory I keep arriving at:

We are, extuitively, sending emotional energies to our past selves, who receive it unknowingly, intuitively.

And, if you didn't know, as defined by the dictionary, to be intuitive, is to 'intuit'. It's the ability of 'intuity'.

Do you see it...

As the dictionary is completely oblivious to, the opposite, to be extuitive, is to 'extuit'. It's the ability of 'extuity'.

That's right.

Extuity.

It's not a word I invented, it's a word that *should* exist, but doesn't.

Weirdly, it's like the English language always knew it would.

It's like reality knew intuition always had an antonym, an opposite.

It just works, and so beautifully well.

Extuition. Extuitive. Extuitively.

Extuity.

So... at long last...

I present to you now, the Theory of Extuity.

The Theory.

We've covered a lot.

But it's not about the amount of content we've covered, it's all about the impact of the content we've covered.

It's crazy, ridiculous, incomprehensible.

Perfect.

Crazy is what we want. Crazy is exactly what new discoveries start out as. Or, as our dear Albert Einstein puts it, "If at first an idea does not sound absurd, then there is no hope for it."

I absolutely love that quote.

However, 'crazy' and 'absurd' is one thing, but 'incomprehensible' is another. Making this theory *comprehensible* and easier to understand is something we can address.

So, it's time for a quick recap.

Brace yourself, we're going to step back through time...

On Level 1, we uncover the flow of time is the flow of energy, and more specifically, time is energy. As we have experienced with our GPS satellites, any flow of time can be sped up or slowed down, a phenomenon called "Time Dilation". It was just as Einstein proposed in his Theory of Relativity. The greater the energy an object experiences, the slower its flow of time, and the faster time moves around it. Time is therefore not a universal constant, it's different for every object in the universe. Furthermore, each unique flow of time is not constant either; its rate of flow is ever-changing. Finally, we introduce "timewaves", a new model of a timeline that represents time as energy.

LEVEL 5

On Level 2, we dive into Quantum Physics, "Quantum" literally meaning 'energy'. On this level, we discover how the "Double-Slit Experiment" reveals energies can be in all possible energy states at the same time until observed; "Quantum Superposition". Applying this phenomenon to the "Quantum Cat", we evolve the timewave model into a "Time Helix", representing how one entity can be on multiple timewaves at the same time. With superposition, we discover how energies can be synchronised or 'entangled' across space, known as "Quantum Entanglement". More profoundly, it also proves energies can be synchronised across time with a clear stance on it being directionless; the energies could be sent both forwards and backwards through time.

On Level 3, we investigate the connection between emotions and our bioelectricity. Understanding emotions are labels for our energetic states, that emotions are energies, we explore how we experience time in differing states of emotion. Outlining how states of high emotion make time seem to fly by and how states of low emotion make time seem to slow down, we unveil a startling connection to Time Dilation which we call "Biological Time Dilation". We highlight the connection that emotions have with time and explored the "Energetic Body" counterpart to our physical human body, our "Quantum Body".

On Level 4, we dive into another phenomenon of the human experience. We uncover how "intuition" – an accepted mystery of humanity – is an experience that directly involves emotions. It specifically occurs when emotions transcend space or time. There's a clear connection between the characteristics of intuition and

Quantum Entanglement, leading to the proposal of "Biological Quantum Entanglement". Our focus, the time aspect of it, was validated by a research study by the HeartMath Institute. Before diving into the implications of this, we are taken to Level 5.

This level.

On Level 5, we uncover that intuition – the ability to receive emotional energies across time – has a counterpart. By definition, anything that is 'received' must be 'sent'. This means emotional energies are sent across time, the opposite of intuition. There is no word for it, but exploring a possible 'opposite' leads to the word "extuition". As "intuity" is the word defining one's ability to intuit, or be intuitive, "extuity" is the word defining one's ability to extuit, or be extuitive.

And here we are.

The Theory of Extuity.

However, one thing needs to be made very clear. The "Theory of Extuity" isn't just the specific ability defined by extuition. The Theory of Extuity is everything that makes extuition possible.

The Theory of Extuity is a theory that is built entirely upon a new understanding of time. It's based on the energetic understanding of emotion. It's based on a temporal understanding of intuition.

So, now I'm challenged to officially define it.

Here's my take...

Theory of Extuity: *All emotions radiate outwards from the point in time it is experienced and can be felt at any other point in time, with an energy proportional to the strength of the emotion and inversely proportional to the distance of time between them.*

Perfect. That's exactly it.

For my fellow mathematicians, there's a formula too:

$$E_2 \propto \frac{E_1}{(T_1 - T_2)^2}$$

If you're not a fan of mathematics, I'm not going to spend time on this. All I'll say is that if you're unfamiliar with '$\propto$', it simply means 'proportional to', while T_1 and T_2 represent two points in time, and E_1 and E_2 represent emotional energies at those points in time.

Now, I'd highly recommend you read the written theory once more. As you can see, I've been a bit cheeky.

I've introduced two new aspects of extuity in the definition we haven't covered yet, but two which you may have already noticed.

When we dived into intuition, and if you think about all the times that you've experienced your own intuitive feelings or "gut feelings", there's a very important aspect of it that I didn't point out...

The *type* of emotion that was felt in those moments.

As we saw from Sarah's experience when she lost her mother that many will be able to relate to, the specific emotions were of deep sadness, loss, and depression. It wasn't just a mix of extremely low energy emotions, the emotions were also very strong, felt very deeply. Likewise, as we saw from my experience when a pickup truck drove dangerously across the main road I was driving down, the specific emotions were of deep worry and fear of injury and death. Again, it wasn't just a mix of extremely low energy emotions, the emotions were once again very strong. It's the same with the

HeartMath research study; the emotions that were triggered were intense.

You have to remember that these are all felt unconsciously, but because they are so potent, they are felt to an extent that affects conscious thought or behaviour.

Hence, the *stronger* the emotion, the stronger you can feel it through time. And notice I said 'through' time. As all feel in intuitive moments, the *closer* in time you are to an emotionally triggering event, the stronger the intuition or "gut feeling" of that event. I believe it's this second element that allowed HeartMath to definitively detect temporal intuition in a test environment; they were *very* close in time to an upcoming emotion.

It's these two aspects I added to the definition of the Theory of Extuity.

One, *"proportional to the strength of the emotion"*, meaning the stronger the emotion when it is experienced, the stronger the energy that is felt across time. Two, *"inversely proportional to the distance of time"*, meaning the further in time you are from the event, the weaker the feeling across time.

They are important aspects of extuity to keep in mind.

Now, let's focus on the core part of the Theory of Extuity: *"All emotions radiate outwards from the point in time it is experienced and can be felt at any other point in time."*

Let's revisit the Time Helix model to help us visualise how this works.

LEVEL 5

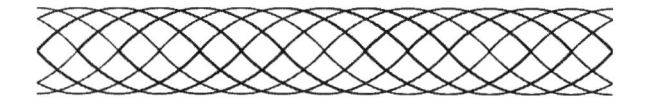

The Time Helix above represents a time period of regulated/balanced emotions. Without strong emotions, extuition isn't as strong, and intuition is more difficult to feel. But as we've covered, the stronger the emotion at a specific point in time, the stronger the extuition from that point, and this is how it would look:

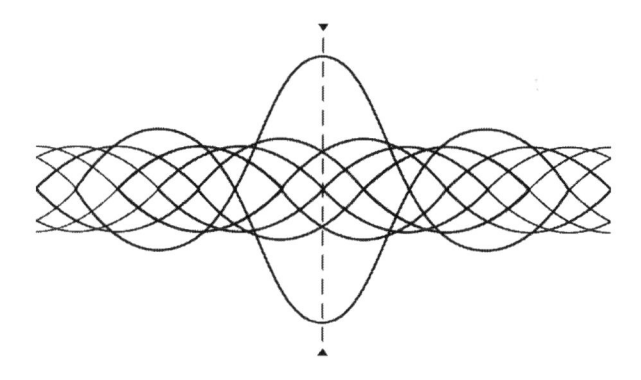

Now we can see how the emotional energy of a high-intensity moment radiates outwards in both directions of time. This is extuition, our emotions being 'sent' or 'emitted' outwards from a point in time. Those energies are 'received' and felt through intuition at other points in time, in the past or future.

To be clear, this highlights how our future selves communicate with our past selves.

What this means for our reality is truly shocking.

What this means for *humanity* is indescribable.

The Puzzle.

There are two sides to the coin of extuity.

This theory doesn't just highlight that our present self is connected to our future selves, it implies that our present self is also connected to our past selves.

Not through memory, but through energy.

I questioned what this meant, if anything. Sometimes powerful secrets can be found in places we least expect. So I thought endlessly of what it meant to be energetically connected to our past and what experiences in life it could apply to. What came to mind was a life-changing human experience we've never had any reason to question…

Grief.

Grief, as you know, is a result of loss. It's one that applies to more events than just the loss of a loved one. It's one that can be just as psychologically severe for the end of a relationship, a friendship, a job, a business, good health, and more. It's a deep emotion caused by the loss of a deep emotional connection.

And what's particularly interesting is what people say to those battling grief…

"It gets better with time", or "Time heals all wounds".

This theory adds an entirely new understanding to these accepted yet incomplete truths.

Because of our ability to hold memories of the past, humanity has assumed that all emotions related to an event in the past are only felt in the present because of our memory. We believe "It gets better with time" as our ever-changing lives provide more distraction from the grief, but if you've experienced strong grief, you'll realise there's another element in play.

As the years pass, consciously thinking back on the loss becomes less and less painful. The emotions we feel are significantly less paralysing. And guess what, the Theory of Extuity provides the perfect explanation for it.

Let's revisit the theory...

"All emotions radiate outwards from the point in time it is experienced and can be felt at any other point in time, with an energy proportional to the strength of the emotion and inversely proportional to the distance of time between them."

If, as this theory suggests, emotions radiate outwards from the point in time it was experienced, it means we would 'feel' the pain of grief less as the distance of time increases from that event.

The emotional energies passing through time would weaken as the years pass.

And it does.

The theory fits, and not just "kind of fits". It fits to absolute perfection.

It's a beautiful level of understanding of grief. While time doesn't really 'heal' emotional wounds, emotional pain does indeed get better with time. Regardless of how many distractions your life provides from the pain, it will inevitably be felt less and less, especially as you allow yourself to experience other emotions over time as it dissipates the emotional energies bleeding through time from the past.

I don't wish to dive too deeply into the health side of emotions, but I'd like to point out something very important.

As we know, emotions have a physical energetic side to them. This is why we cannot just process them mentally. They must be allowed to be physically processed too. Hence tears when we are sad, screams when we are in pain, laughter when we are joyful. We allow the energies to be released.

And the Theory of Extuity highlights an important aspect of physically processing emotions.

Because emotional energies can be physically felt through time, it's not enough to allow ourselves to physically process the emotions only at the point in time they were experienced. You must allow yourself to continuously process and release those emotions felt by your body for a fair amount of time after an emotionally intense event, and once again, grief is a powerful example of that.

It's extremely important to give ourselves time to process the pain of grief. The emotional energies bleeding through time need a physical outlet. If you ignore the energy your body is feeling for days, weeks, months or even years after a loss, if you stem its flow, your body will experience dis-ease, or as we call it today, 'disease'. Your health will be severely compromised, which is unfortunately a *very* common experience for those going through grief.

Allow yourself to cry. Allow yourself to exercise and release the energy. Allow yourself to process the energies your body is physically feeling through time. Not just in the moment of loss, but for as long as you feel the emotions after the painful event.

And rest assured, over time, the pain will be *felt* less and less.

Exactly as the Theory of Extuity proposes.

However...

The real power of extuity doesn't lie in understanding how our past affects our future, it's how our future affects our past. It's how our future is connected to our past and present.

It's reality-shifting. Mind-bending. Life-changing.

And, holy sh*t, are the implications *unbelievable*.

Which, finally, brings us to Level 6...

Side-Level 5

Past Lives?

Heard of "Reincarnation"?

Up to 50% of the global population believe in it, that we are reborn into a new life after we die. Or in other words, that we have lived past lives and we will live future lives.

And that intrigues me.

It's an aspect of time that is beyond the limits of our 'current' life, yet one that the Theory of Extuity could add an interesting new angle to…

Is it possible that extuition can be felt from a point in time beyond our current lifespan? Thereby potentially proving that past and future lives exist for us?

An interesting question came to mind that could point to this possibility, and I pose it to you now…

Have you ever felt an illogical yet unbelievably strong emotional connection to a specific historical period, event, place, architecture, music, paintings, or anything else from humanity's past? One that you just can't explain?

What's especially intriguing is that there was a word coined for this inexplicable experience because of how common it is. "Anemoia", meaning: nostalgia for a time you've never known.

If you relate, it might just indicate the impossible, that you did indeed have a past life in that period or related to that specific event or work of art.

Or... as we've covered, those emotions could also originate from your future in this life, from a movie, book, or hobby that your future self will experience with intense emotion.

So, extuition from a past life? Or extuition from your future self in this life?

I'll let your ego decide which one to believe, if any.

Just don't let anyone else decide for you.

SIDE-LEVEL 5

YOUR JOURNEY

Level Progress & Key Learning Points

1. Time is energy
+ Time is elastic + Time is not constant

2. All points in time can be connected
+ One can be on multiple timelines

3. Emotions are energies
+ Biological time is elastic + Quantum bodies exist

4. Emotions transcend time
+ Intuition is receiving energies across time

5. Extuity influences past & future
+ Extuition is sending energies across time

6. 🔒

7. 🔒

Level 6

The Language of Extuity

Navigating infinite futures. Redefining 'fate'. Rewriting pasts.

Crazy. Ridiculous. Incomprehensible.

But *why* is it considered crazy?

Quite simply, it's because it doesn't fit our current understanding of reality. Our subconscious fear of realities that we haven't experienced pushes us to reject the 'crazy' ideas and cling to the reality we know, our survival mechanism.

A survival mechanism that hinders change, especially reality-shifting changes.

And this is where I want to propel you into the future.

Our great ancestors would see our world today as impossible, but we don't have to be in the same boat. We don't have to reject the 'impossible' ideas and wait for our descendants to accept them. We can fly ahead in time.

So, let's do just that. Let's embrace the impossible future, today.

And where do we start?

We start with the *language* of extuity.

Decoding Energy

I'm a programmer.

I don't just code using different programming languages, I design and build pathways between entirely different systems. I ensure different entities can talk to each other. No, that doesn't just involve building communication channels. Sending and receiving data is just one element. I also have to design them with the ability to *understand* each other. That means crafting *how* they can understand each other.

And it's no different with the Theory of Extuity. It's no different with reading our futures. The data stream is there, a communication channel exists, and the foreign language has been identified.

Bioelectrical energy.

And we're very blessed with the natural ability of translating it into a language that we can understand.

Emotions.

The challenge now is understanding how we can *read* this language, the language of emotions, and how we can process it to read specific aspects of our future. But first, in order to read it, we need to know where it's written, and how. We need to understand how it *programs* us, and from there we can identify the specific source code that originates from a source other than our past and present.

And if you've ever studied psychology, our emotions literally program our minds. Our emotions define *every* aspect of our lives. And yes, I really do mean *every single* aspect of our lives.

But not everyone realises it.

A common argument is logic vs emotion. If you think of yourself as a logical being, heads up. Even logic is influenced by emotion. Logically, a glass can be both half empty and half full, but your emotions will direct your logic to a specific perspective. So if you really love your logic, you should now understand how this fact affects every single 'logical' decision you've ever calculated.

Emotions define *every* aspect of our lives. Emotions define who we are. Therefore, it also defines our entire world around us, because we don't see things as they are, we see things as we are.

That's because emotions influence our very thoughts.

We feel more comfortable thinking about certain topics, ideas, or people because they make us feel curious, excited, and alive. Of course, we already know this, but what does this mean in the context of extuity? What does it mean for our thoughts if our subconscious mind is influenced by a data stream of emotions from our future selves?

One result would be our subconscious mind influencing our conscious mind – entirely without our awareness – to focus on a thought of an event before it even occurred. At least, it's what the code of extuity suggests... but does it happen?

Yes! It's the phenomenon we call "Coincidence".

Think about it.

Have you ever found yourself thinking of an event, topic, or situation, and it just happened to occur or show up a few moments later? Thinking or talking about someone, and a few minutes later receive an unexpected message or phone call from them? Or saying it would be funny or crazy if something specific happened, and shockingly it does??

When it happens, we all think the same thing: "What a weird coincidence" or "That's crazy!"

Well, it isn't a "coincidence" or a "synchronicity", and it definitely isn't a "manifestation". Your emotions pulled your mind to the thought of it because your emotions a few moments in the future were entirely focused on that person. Your future self influenced the thoughts of your past self through the energy of the emotions bleeding through the short distance of time from that event.

Coincidence is an illusion. It's a side-effect of extuity. It's subconscious communication with your future self.

Oh, and if you are a spiritual soul, "Manifestation" and the "Law of Attraction" are illusions too. To be clear, I'm not saying they're impossible, they're just misunderstood. However, that requires us to dive a little deeper into extuity before we can begin to understand how they really work.

Now, identifying these unexpected subconscious influences from our future self is helpful to be aware of as it leads to fewer moments of shock and confusion, but it's not something we can really use to 'read' the future. It influences our minds in such a short span of time, we don't have much time to analyse and use the information.

Which led me to asking the most important question of all...

What deep elements of our mind have *already* been influenced by future emotions? What aspect of our psychology can be queried to identify specific data linked to our future? Do any even exist?

Yes. Yes, they do.

If you're a psychologist, you've probably already tied 'emotions' to what is possibly the greatest factor in what defines our very being...

Our opinions.

Opinions define our personality. Opinions define our values and principles. Opinions are the aspect of our psychology that few realise are connected directly to our emotions.

Think about it.

What is an 'opinion'?

Let's look at a few examples. Hiking is fun. Politics is boring. Music is beautiful. Science is confusing. Travelling is magical. Wars are horrific. Life is a gift.

Opinions are statements of how we *feel* about something specific.

An opinion is an *emotion* towards something specific.

Opinions, I discovered, are the key to reading the language of extuity, of decoding the communication from our future selves. I'd like to introduce you to its power by highlighting a more influential phenomenon that we experience more frequently than we realise.

Have you ever had an immediate opinion about something or someone upon first interaction? An opinion that made no logical sense as to how your past experiences would arrive at that conclusion?

One of my favourite examples is "Love at first sight".

And no, I'm not just referring to romantic love. But yes, we are about to reveal the truth behind this romantic idea.

Maybe it was a sport, a game, a country, a business idea, a new food dish, or, yes, even a person.

THE THEORY OF EXTUITY

Have you stumbled across something or someone that you felt inexplicably drawn to? An unexpected feeling of great excitement, love or deep curiosity rising up within? And when you experienced it, you thought "I knew it!", despite not having any prior experience of it at all. There was absolutely no logical reasoning behind your 'knowing', and you may have even been slightly confused at how you knew.

Think long enough, and you'll find countless examples.

This was your opinion coming from your future self.

This was you unconsciously tapping into your future energetic emotions towards something specific.

It's a way we can communicate with our future selves, analysing our opinions moment to moment, topic to topic, place to place. It's how we can *read* the energy, by questioning our emotions towards something specific.

A key part of this is understanding that the more we experience something throughout our future lives, the stronger our opinion of it in the past and present. Why? Because those emotions are being emitted from even more points in time, so will naturally be energetically stronger. It's why "love at first sight" is such a common phrase in human society. Love for a romantic partner is one of the most intense emotions we can feel, and one we experience for a long period of time.

As you may have experienced yourself when you first met your partner, you had this odd feeling like you already knew them. This also applies to meeting a life-long friend or close colleague for the first time.

That's because your future self is extremely familiar with every aspect of their personality. Your future self has spent so much time with them, has shared so many experiences and emotions with them, your past self unconsciously feels these emotions. Your past self has been subconsciously programmed by your future self with opinions that make you feel like you 'know' this person without having met them before.

Your opinions of them, your emotions, are extuitively transcending time.

By being aware of these inexplicable opinions and emotions, you can begin to 'communicate' with your future self. You can begin to 'read' future energies. The more sensitive you become to it and the more in tune you get, the stronger your ability to connect with your future self.

Once again, this doesn't just apply to people. This applies to anything and everything you experience that you have strong emotions for throughout your life; sports, places, ideas, food, hobbies...

The list is endless.

And remember how opinions define us? They define our values and principles. They define our personalities. This highlights a very interesting revelation for humanity.

As parents know very well, regardless of how multiple children are raised in the same environment, even if near-identical with the same homelife and homeschooling, their personalities are drastically different. This is because our minds, thoughts, and opinions are not just influenced by our pasts. Our emotions and opinions are unconsciously felt from our future selves, hence partially programmed by our futures.

And there's a reason this is most noticeable in children.

Children have little to no 'past' to program any deep opinions or personalities, and regardless of how similar childhoods are for siblings, their futures are never the same. They'll end up with different career paths, different partners, different opportunities. Entirely different futures, hence significantly different emotions and opinions throughout life. Their drastically different future selves are writing their personalities from a young age.

Their most-likely futures are writing their pasts.

And as we highlighted earlier, this isn't just a phenomenon that applies to our childhood. Our futures are consistently writing our past and present. That means in this very moment, your future self is, by definition of extuity, sending back emotional energies that you are not consciously aware of. Your future self is influencing every opinion you have about every aspect of your life, including every aspect of your very personality.

And you can use it.

You can learn to read it before it writes your story. You can learn to read it and choose how to let it direct your entire reality. It's the subtle language that our future selves are using to communicate subconsciously with our past selves. It's a subtle ink that our future selves are using to write our paths.

You can use it to read your infinite futures, and rewrite it, reshape it, redirect it.

And to truly understand *how* to read this language and *how* to use it, it's time to unveil the most powerful aspect of intuition and extuition.

Time Compass

Why is it that we can *feel* emotions, such as the pain of grief, so much stronger from our past than our future?

Think about it for a minute...

Here's another question that can help: Why is it that intuitive or gut feelings are strengthened when we see the very thing they are related to?

Really think about it...

The answer was actually revealed to me when I dived deeper into 'Opinions'. Remember, an opinion is an emotion towards something specific. An opinion is *felt* when we *focus* on something specific. An emotion is *felt* when our conscious mind *focuses* on something specific.

Do you see it now?

Our conscious mind, our focus, acts as filters and amplifiers for our emotional energies. By simply 'thinking' or 'focusing' on a particular idea, person, event, or circumstance, we energetically connect to the emotions tied to it. It's the secret to reading specific energies received from our infinite futures.

Our consciousness can amplify our energetic connections through time.

Grief is a perfect example.

Soon after a loss, when we are distracted, we still feel the emotions of the loss bleeding through time. However, when we consciously think of the loss, those emotions are amplified. When we consciously focus on the specific event that triggered the emotion, we amplify the emotion bleeding through time.

It's the same with emotions from our future.

When I was driving in Wales, I felt uneasy, and it was only when I focused on the junction ahead that I was able to get a clearer 'feeling' of the impending danger. Likewise with "love at first sight", it's only when we see the person, when our conscious mind focuses on them, that we feel emotions from our future relating to them.

And it's the same with any intuitive or gut feeling. In fact, there's a really funny example of this, and one you should definitely be able to relate to...

When you suddenly feel like you've forgotten something.

You may have just left the house, packed your bags for a trip, or left a store. As you walk out, you have a feeling that you've forgotten something important. So, what do you do? You think about all the possible items you could have forgotten, but with no luck.

And guess what, 30 minutes later when you open your bag to pull your keys out, or wallet, or phone, you don't find it there. You left it behind. And what do you do? You think back to the moment you left the house, or the moment you finished packing your bag, and you wish you'd remembered it.

This is an example of extuity, one where your future self is sending emotions of regret and frustration to your past self. In the moment you felt you forgot something, when you focused on the possibility of having forgotten something, that feeling would have gotten stronger, and you may have also noticed it getting stronger and stronger up until the point you discovered what it was.

Once again, conscious focus can amplify our energetic connections through time.

It's really important that you understand this. Our conscious mind is the key to amplifying energies that we feel through time. It's the key that can be used to open doors through time.

And now it's time to reveal a new aspect of time. Specifically a new dimension of time. And to do so, we are going to revisit a story that we covered right at the beginning of this journey.

My journey, the journey of impossibility.

In 2012, I received an offer to study at the University of Cambridge, but, as you know, I wanted to go to Imperial College London to study their more practical Computer Science course. I was torn, I didn't know which offer to accept, which choice to make.

In 2014, I wanted to drop out of university and start a company. Everyone was telling me *not* to, to stay and get a degree. Again, I was torn. I didn't know which choice to make.

In 2015, after building a company valued over £3 million, I felt like the best route for me was to leave and work on more software-focused companies. Money vs happiness. I was torn, I didn't know which choice to make.

I know you can relate.

I know you yourself have hit crossroads in life, faced with difficult choices that led to completely different lives. I know you have, because we all do. It's a part of human life, but more importantly, it's a part of 'time'.

They are called "Choice Points".

Choice points are points in time where we consciously decide between two paths, two completely different paths through time.

They each have their own emotions, their own energies, their own timewaves.

What makes them so difficult to choose between is often a case of logic versus emotion. Logically, according to modern society, choosing to study at the University of Cambridge was the sensible choice. Logically, getting a degree was the intelligent choice. Logically, continuing to work at a self-made company valued in the millions was the common-sense thing to do.

But logic isn't the key here. Logic is the false compass used to navigate paths of time.

The key is extuity. The key is energies felt on each time path. The key is *conscious energy*.

Somehow, before I even knew of extuition, I used the power of extuity to help me navigate each and every one of those major choice points. Somehow, each time, I thought to ask myself a very powerful question...

"50 years from now, which decision would I regret?"

I tied together time and emotions. My past self emotively connected to the future self I wished to become. I envisioned myself living a life of pure freedom, having built companies with products that changed the world, having built charities with projects that saved lives. I consciously focused on a point in time that I wished to reach, amplifying its energies, and I then focused on each choice before me.

In 2012, I thought about the future of studying at the University of Cambridge, and I thought about the future of studying at Imperial

College London. I felt indescribably strong gut feelings to take the Imperial College route. It was so strong, it outweighed the logic.

In 2014, I thought about the future of staying at university for 2 more years of project work, and I thought about dropping out and starting a company. I felt the joy, the nerves, the excitement and adventure of building a company. When I consciously thought "degree", I felt absolutely nothing.

In 2015, I thought about continuing to build the company and potentially make millions, and I thought about leaving to take everything I'd learnt on a route of happiness. I labelled it as a route of 'happiness' because it was the overwhelming emotion I felt. I took it, and I'm living the life I dreamt of.

In each and every moment, at every choice point, I allowed myself to focus on each path, I allowed myself to subconsciously feel the emotions of each route, I allowed my mind to connect to each timewave of energy, each Time Helix, through time.

In each moment, I connected to each future, and through conscious focus I amplified the energies that my future selves on each path were extuitively sending back in time.

This isn't fiction. This is a real life, a real story, a reality that defied modern logic because I navigated with intuition *because* of extuition.

And of course, I'm not the first or only human to do this.

In fact, the *most successful* people in history across completely different paths in life have stated their *heavy* dependence on "intuition", but because it's not understood by society, we don't hear about their praise of it. Who? Steve Jobs, Cillian Murphey, Conor

McGregor, and even our dear friend in science, Albert Einstein, who said "The intuitive mind is a sacred gift and the rational mind is a faithful servant."

Steve Jobs praised it so much, he stated "Intuition is a very powerful thing, more powerful than intellect."

Connecting to extuitive energies from our future self is the secret key to navigating time.

I call it the "Time Compass".

To understand this better, let's have a look at the traditional model for dimensions of time:

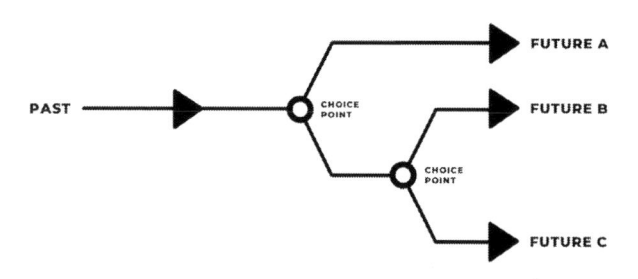

```
PAST ─────────▶ ○ CHOICE ─────▶ FUTURE A
                 POINT
                      └──▶ FUTURE B
                         ○ CHOICE
                           POINT
                             └──▶ FUTURE C
```

At every choice point, there are multiple paths that can be taken. Each path through time is a different reality. Each path through time is a different timewave of energies.

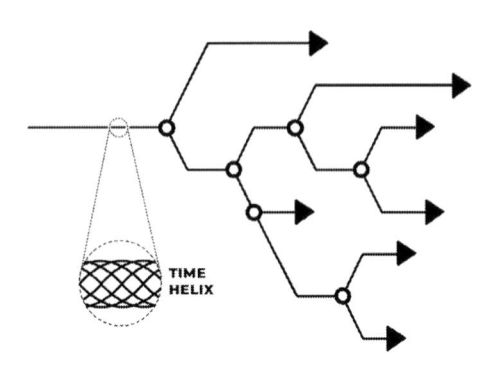

[Figure: Time Map diagram with branching arrows emerging from a "Time Helix"]

I call this a Time Map. Everyone has one, and it outlines all of our potential paths through time.

Now, when we hit a choice point in life, the trick to identifying which path is best is down to where we direct our conscious thought in order to filter the extuitive energies on a specific path. For example, when I was deciding whether to drop out of university, I allowed myself to focus on each choice. Every time I considered staying at university and getting a degree, I felt nothing. But *every* time I considered leaving and starting a company, I had that powerful 'gut feeling' shouting "YES!".

I allowed myself to 'feel' what my future self would feel on each path, really connecting with my gut feeling, and that's how I 'knew' which was the best one for me.

Now, here's something really important...

At any choice point in life, asking the *right* question is key. You cannot simply ask "Which is best for me". Time doesn't work like that. Extuity doesn't work like that. It's *binary*. You need to ask a Yes or No question. You need to think "Will I enjoy option A", or "Will I enjoy option B", and *feel* which one you react emotionally stronger to.

You can't ask "What is my purpose". You need to ask "Is my purpose on this path".

What you'll receive in response is a binary gut feeling, a high or low emotive energy, a "Yes" or "No". If you feel excited, curious, or inexplicably drawn to it, it's a *Yes*. If you feel something is off, a sense of nervousness, worry or fear, it's a *No*. The stronger the feeling, the stronger the answer.

It's significantly easier to do than you think.

Where it gets difficult is when your logical mind interferes. It's not always clear why a choice may be better for you, and it's crucial not to let a lack of logical understanding stop you from doing what you 'know' or what you 'feel' to be right.

There is something else you should be aware of.

Just because a path is better for you, one that gets you closer to a life living up to your greatest energetic potential, it doesn't mean it'll be *easy*. Remember, I broke my arm snowboarding in 2023, but it didn't mean I wasn't on the best path. In fact, it forced me to take time off work to reconsider everything I thought important, and it led me to great discoveries I would not have otherwise made. I was on the path through time that led to me living one of my most impactful lives.

More importantly, however, is the *obstacle* to mastering the ability of navigating through your Time Map.

THE THEORY OF EXTUITY

Your ego.

Ancient Teachings

Now, don't fall for the common misconception that the 'ego' is bad. Your ego is an integral part of who you are. Your ego fuels your ambition and defines your identity. It's only when it's left unchecked without any self-awareness that it can cause conflict.

But unfortunately, it can also cause a problem for navigating time.

The problem lies in the ego's role as a survival mechanism.

As mentioned a few times already, our survival traits can hinder change, especially reality-shifting changes. Any potential new reality is a world our ego does not know it can survive, so it will try its best to deny it, to reject it, in order to secure our survival.

This is particularly disruptive when it comes to choice points in time.

Our survival mechanisms are built upon our fears, or rather, our fears create our survival instincts. Fears generally help us avoid perceived dangers. Our survival traits are also built on desires for respect and fame. Through this, we gain recognition and respect from our 'tribes' and we're less likely to be kicked out to survive on our own.

Our egos are built on a foundation of emotions. These emotions, which left unchecked, can greatly influence and distort the emotions we intuitively feel through time.

Let's revisit my decision to drop out of university.

Most people would never have considered dropping out, regardless of how big a pull they felt to other opportunities. They would have been pushed away from it by their ego's fear. Their fear of being thought 'idiotic' would have on a very deep level kickstarted their survival centre to stick to the path everyone respected, ensuring their respect from their family and friends, ensuring their place in their 'tribe'.

Likewise, my decision to reject the offer from the University of Cambridge is an example of the opposite nature.

Most people would have snatched up that offer without any thought. They would have been pulled to it by their ego's desire. We are brought up in a society where students of Cambridge, Oxford, Harvard etc. are respected, honoured, held up high on pedestals. It's gold for our egos. It wouldn't have mattered if Imperial's more practical computing course had been better for them too. The emotion their ego would have evoked would have completely blinded them to the better path in life, the better route through time.

In these cases, you should question *why* you would make a choice you are *pulled* to; is it ultimately because of how someone else will feel? If so, your ego is in play, and it may not be the best choice for you. Just as important is asking why you would *not* make the other choice. That's for the situation where your ego is *pushing* you away from a path.

However, that's unfortunately not the greatest obstacle to overcome.

On a deeper level, our egos have the power to manipulate the intuitive energies we feel, or rather, *how* we feel about them. It will keep us on paths of pain even if there might be a path of light right in front of us, and it comes back to our survival traits again. I heard

LEVEL 6

someone describe this in a very powerful way: "Our nervous system will always choose a familiar hell over an unfamiliar heaven."

It's as I mentioned earlier.

Our ego will keep us in the reality we know, rejecting new realities we don't. It *knows* it can survive in the familiar, but doesn't *know* it can survive the unfamiliar. It's a very powerful truth to understand because it affects many goals in life.

Without realising it, for example, even if you crave a life of wealth, you will subconsciously make choices that will keep you on the more 'familiar' paths through time where you are financially poor because your ego knows you can survive that life. In this case, the trick to helping you navigate to a life of wealth is to remove all self-identification with your wealth status. Stop saying, thinking, and believing you are poor. Remove any emotional feelings towards it, otherwise you will subconsciously stay on the path where those emotions will remain true, where your ego knows you can survive. Your goal is for your ego to no longer see the 'poor' paths through life as 'familiar'.

In order to master time, master your mind and ego in the present.

Which takes us to the teachings of monks and gurus.

In these ancient cultures, the very first thing they are taught to do on their journey of achieving self-mastery is to master their emotions. Many believe the purpose of this first step is purely to become more mindful and achieve inner peace.

Everyone's missed the point.

Few know that in these ancient cultures there are levels to the teachings, levels to the abilities they are taught. They must master the first teachings in order to progress to the second, and so on. They actually forbid even mentioning the more powerful abilities to those on the earlier levels. Today, many of these higher abilities are completely alien to us, abilities we consider pure fiction, fake.

And guess what... every single one of them is dependent on mastering emotions. Not for mental health, but because of the power behind emotions, the power behind our biological energy.

While I haven't yet come across any advanced ancient teachings related to time, I believe the Theory of Extuity is one of them, and one of the highest ones. I believe the ability to navigate through time is one of the most advanced abilities humanity possesses. It follows the ancient rules as it too requires mastery over your emotions.

If you don't master your emotions, you have no hope of mastering time.

If you don't understand and tame your ego, your fears and desires, you will be pulled down paths of time that don't serve your greatest potential. You'll have no control over your 'fate'. You'll have no ability of writing your own future.

Once you master your emotions, once you master emotional intelligence, you'll be able to identify which emotions your future self is sending extuitively through time. You'll be able to discover the best paths to take at every choice point in your life, being able to sense the right answer to the binary questions you ask, feel extuitive energies over those from your ego, and navigate through the many futures ahead of you to the one where you live and experience your greatest energy potential.

This, by the way, is the true definition of "Manifestation".

If you're unfamiliar with the concept, many believe that they can "Manifest" the future they desire by simply believing in it, by simply manifesting it with thought. That's not the case. It's why many discover that they constantly fail to 'manifest' what they wish for.

It's also the true definition for the illusion of the "Law of Attraction", which is the belief that what you think or feel is what you attract. I describe it as an illusion because it's more accurately the "Law of Navigation".

However, in both cases, in order to manifest or attract the life you wish, you need to master extuity.

You need to master your emotions in order to prevent your ego from inducing bioelectrical states that blind you to your intuitive energy, blinding you to the energies you feel through time. You need to learn to ask the right questions at each choice point, connecting you to the most informative energies across time.

It's how you use your internal "Time Compass".

And now, prepare yourself... because it's time to reveal the most powerful way one can rewrite their *entire* future... a power very few realise they have...

You can *create* choice points.

Yes. You can *consciously* create choice points. You can create entirely *new* futures. You can redirect and reshape your *entire* life trajectory.

This, by the way, is where logic plays a powerful role.

It's time to revisit a different part of my story.

In 2023, shortly after I broke my arm snowboarding in the French Alps, I refused to let a 6-month recovery period being physically handicapped prevent me from growing and experiencing the wonders of the world. One of the many journeys I embarked on, including this very one that led to the discovery of extuity, began by asking some very powerful questions:

"What do I believe with absolute certainty? What pre-conditioned facts exist in my mind, in my view of reality, that I've consciously or unconsciously accepted on my journey through life?"

Don't underestimate the difficulty of these questions.

Answers could include long-standing beliefs and accepted facts of reality such as *"I need a good salary to live a good life", "I hate spiders", "I can only live in the country I was born in", "I'm not creative", "I believe in God",* or *"Time travel is impossible".*

Hopefully you can see how questioning your deepest views of life can create choice points. It might seem obvious to say this, but we only ever question what we don't have answers to. So, if we let go of an answer we believe is set in stone, we now have a question. We now have a new choice point. We now have a new potential future.

Our logical minds have the power to decode deep emotional programming and create entirely new futures.

Questioning each of the above could translate to *"Is there a better alternative to my job?", "Could I love spiders?", "Where could I live abroad?", "How could I be creative?", "Why do I believe in God?",* or *"Can I prove time travel?".*

The bigger the new question, the greater the new choice point, and the more profoundly different that new future will be to any future you currently have ahead of you.

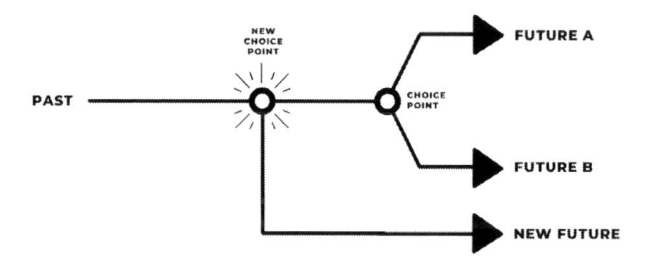

Remember, you don't *have* to change any beliefs. You don't *have* to take a new future path. What's powerful here is allowing yourself to *consciously* consider them rather than subconsciously living with them, the subconscious programming that has defined your entire life trajectory.

By questioning *everything*, you'll find out if there are any beliefs holding you back from your greatest future potential.

As I'm sure you can understand, mastering your ego and hardwired emotions is absolutely vital for this. Mastering your fears and desires instead of allowing them to define you and your deepest beliefs is what opens up entirely new futures. Preventing your ego from blinding your senses to extuitive energies unlocks the ability to navigate through time on a level that many believe impossible.

Upon asking a new question without ego, fear and desire influencing your answer, you'll allow your consciousness to connect to new future paths and the energies you'd experience from them. You'll be able to *feel* the energies of the new future path, and you'll know whether that future is closer to the life you wish to live.

Which brings us to a direction for your compass.

Like all compasses and all journeys in life, you need a destination. You need a goal.

Since my early teens, I've always had a vision for my future self. It's like a "Vision Board", but one I've always seen in my mind. I can clearly see myself living a life of pure freedom, freedom from the constraints of time, location and money. I own companies with impactful products, the profits of which are fuelling charities I've created that are saving real lives. I have a beautiful, supportive, adventurous family, living in a truly beautiful country.

My vision, my goal, is the guide for the compass I use to navigate through time.

I've always been able to 'connect' to my future self at whatever point in time I've reached that goal. I can truly feel all the emotions I know I will feel at that time. Oddly, I can also feel all the difficulties, blessings, ups and downs that led to it.

When I hit each of the three difficult choice points in my teens, I used this vision as a direction for my life.

A compass is useless without a destination, so outline your vision.

It's the very reason Vision Boards work.

So if you feel it will help, put together a Vision Board for yourself, images that represent the future you wish for your future self. Once you see it, feel *every* emotion that you'd feel if you were there. Remember it. Embrace it.

That, by that way, is absolutely crucial.

It's not just about the 'vision'.

LEVEL 6

Embodying the emotions you'd feel at that future point does something quite spectacular in the present moment when you hit a choice point. Remember, your Time Helix contains near-infinite timewave energies. By embodying or 'synchronising' with the emotional energies of a future point in time, you'll be 'pulled' towards the deep energies inside your Time Helix that have the potential to materialise that future reality, hence 'pull' you towards the paths at choice points that lead to that future energetic state.

It's a secret to speeding up your journey through time to the future you wish to experience in the here and now, allowing you to find the shortest route there.

So when you hit choice points in life, keep your target goal in mind and target emotions in your heart. One by one, focus on each path through time ahead of you, ask the right questions, and feel the energetic heartfelt pull. If you feel you're stuck in time, create choice points by questioning your strongest beliefs of reality. Keep your ego in check, and allow the energies of your future self to reach you. Feel it in your heart. Feel it in your gut. Read the language of extuity, the emotions towards the specific choice, and transcend time.

This, by the way, is the answer to one of humanity's greatest questions.

Are our fates set in stone?

Does "free will" truly exist?

Can we live the life we wish to? Or is our destiny already decided for us?

Unfortunately, fate is indeed set in stone for the majority of humanity, for those who do not know how to navigate their Time

Maps. For those whose egos are too strong and whose emotions cloud their extuitive energies, the decisions they make at choice points are pre-programmed. They would *never* have taken the alternative route, meaning the choice points didn't actually exist. And without choice points, there is only one path.

Moreover, without ever questioning their deepest opinions, values and beliefs, which all would have once again been pre-programmed by past experiences, they would never be able to create new choice points as they travel through time, severely limiting the potential futures they could live.

So if there is only one path, and there is no creation of any alternate paths, their path through time is already written.

Their "free will" is just an illusion.

Their fate is set in stone.

However, if you follow everything we've covered, if you prevent your ego from controlling every choice you make and if you master the creation of new choice points by questioning and decoding your pre-programmed mind, you will have the ability to redirect your path through time. You will have the ability to rewrite your future.

It's not easy, but nothing worth doing ever is.

Your fate is only predetermined if you remain unconscious of your connection to time.

You *can* rewrite your destiny if you master navigating through time with your Time Compass.

You *can* write your own fate.

Whether you do is now up to you.

LEVEL 6

Side-Level 6

A Time Hack

I believe I've discovered a "Time Hack".

It's possible to speed up learning a new skill or hobby, especially when you believe you can't.

When you are learning a new technique, a new move for a sport, or just generally anything new, think of and embrace this statement:

"I've already done it."

And as you do, imagine how you will feel when you have done it, and embrace those emotions in the here and now.

What you'll be doing is hacking time.

You'll be connecting to a version of your future self that has succeeded in learning or achieving that specific skill or technique. Pulling those emotions sent extuitively by your future self into the present moment will guide your thoughts and actions to mastering it much quicker.

If you give yourself time to understand everything on Level 6, you'll understand exactly how this works.

And trust me, it works.

YOUR JOURNEY
Level Progress & Key Learning Points

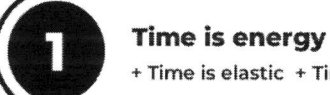

1. Time is energy
+ Time is elastic + Time is not constant

2. All points in time can be connected
+ One can be on multiple timelines

3. Emotions are energies
+ Biological time is elastic + Quantum bodies exist

4. Emotions transcend time
+ Intuition is receiving energies across time

5. Extuity influences past & future
+ Extuition is sending energies across time

6. One can write futures
+ Choice points create multiple futures

7. 🔒

Level 7

Humanity's Greatest Secret

Brace yourself.

Everything we've covered up to this point has been child's play compared to what I'm about to reveal.

In fact, if you haven't already, please take a break.

Go for a walk, make a cup of tea, or just sit back for a while. Give yourself some time to process everything you've read.

I'm about to reveal an aspect of extuity that I'm not sure any human has ever considered before... in fact, in all honesty, even I'm still trying to process it. It's such a mind-bending shift to the understanding of our reality, I *really* hope someone out there can help me understand the vast implications of it.

I'm serious.

Are you ready?

I hope so...

Well, let's do this then.

Untapped Powers

For the past few years now, I've been exploring our greatest pool of untapped human potential.

Becoming conscious of unconscious abilities.

We, our minds and bodies, do many things unconsciously. In fact, you're doing something right now that you're not currently aware of, something you're not consciously controlling, and when I say what it is, you'll think, "Obviously."

What is it?

Quite simply, you are breathing.

Obviously.

To be specific, you *were* unconsciously breathing.

Now that I've brought your attention to it, you are now conscious of it and you can manipulate it. If you wish, you can now breathe deeper, longer, slower.

And here's the fascinating part. If there's one thing that you are doing unconsciously, there are more.

So, what else is there?

Well, we covered one of those things earlier: regulating your bioelectricity. We've been raised in a society that has lost many ancient teachings, including that of understanding, cultivating and applying our bioelectricity like we can with our breathing.

I'm not just talking about emotions being labels for our bioelectrical states. Our bioelectricity goes far beyond our emotions.

As a little hint, are you familiar with the feeling of goosebumps? That sensation when your entire body tingles? Well, that's your bioelectrical energy, or 'Chi' as Chinese Medicine calls it, and it's all unconscious. That tingling feeling in your nose when you sneeze? Bioelectricity. The pain you feel when you get a bruise? Bioelectricity.

You haven't been taught how to consciously control it and move it around your body at will.

Right now, at this very moment, your body is unconsciously controlling your bioelectricity and you are completely unaware of it, just like you were unaware of your breathing just a few moments ago.

What else are we unconscious of?

Here's a really cool one...

Conscious dreaming. Or, as it's known, "Lucid Dreaming".

Have you ever watched Inception? The movie by Christopher Nolan? If you look into the origins of the story, you'll find out it's based on a real human ability, Lucid Dreaming.

When we dream, we have no idea we're in a dream until we wake up. Only then do we realise it was a dream. But while we're in a dream, we have no control over it. We watch it play out. It's a world created purely by our subconscious and played out by our subconscious mind.

This is the first of three types of dreams.

The second type of dream is when you have *some* level of control. You can make decisions in the dream, in the story being played out, or at least it feels like you're making choices. But again, you have no idea it's a dream *until* you wake up.

And the third?

The third type of dream is when you're in a dream, but you *know* you're in a dream while you're in it. You become aware that you are in a dream world and not the real world. You become 'lucid', aware, conscious. The level of control over this subconscious world is down to two things: Your subconscious self-confidence and your imagination. With practice, you can learn to manipulate anything and everything and completely break the laws of physics. *Anything* is possible. Just by thinking it, you can move or transform objects, reverse gravity, alter the environment, and, my favourite, speed up, slow down, or reverse time across an environment or through a single object, such as a plant.

It's yet another ability we are unconscious of.

What's my point?

My point is that there are *multiple* aspects of our human existence that we are not consciously aware of. It's by far one of the *greatest* limitations of our human potential.

Did you know "Conscious Breathing" isn't just capable of influencing our state of mind, like when we take a deep breath to calm down, but it's also one of the most powerful ways of boosting our health. From Chest Breathing, to Abdominal Breathing, to Reverse Abdominal Breathing, we can use different breathing techniques as an "Internal Massage" for our various organs, and in combination with various upper-body poses, we can use it to increase or decrease blood flow to specific organs.

Moreover, "Conscious Bioelectricity" enables one to reduce pain, headaches, motion sickness, and even speed up the healing of sore muscles, cuts, bruises, scrapes and more. Its effects also apply to any unconscious 'tingling' sensation, which funnily enough means you can even use it to prevent a sneeze, or consciously trigger one. When you really learn how to cultivate and control it, you can even consciously create that 'goosebumps' feeling throughout your entire body.

And regarding "Conscious Dreaming", or Lucid Dreaming, when you realise that it's a conscious doorway into your subconscious, the possibilities are extremely powerful. You can boost creativity, increase self-confidence, reduce deep anxiety, and even treat PTSD. It makes sense. Lucid Dreams effectively give you the ability to 'reprogram' your subconscious mind. It can even boost the effects of "Conscious Bioelectricity", specifically healing. That also makes sense, as the brainwave state you're in when you're dreaming, 'Delta' waves, is the same state as comas, and of course we use medically-induced comas for improved healing.

Becoming conscious of unconscious abilities is probably the most powerful thing you can do.

And, to my utter astonishment, I realised this also applied to extuity.

Let me explain...

Intuition is receiving emotional energies across time. Extuition is sending emotional energies across time.

Once again...

Extuition is sending emotional energies across time.

In other words...

Extuition is *unconscious*. We're not *consciously* sending emotional energies across time.

THE THEORY OF EXTUITY

But what if we did? What if we were able to control the energies we send across time?

Well, quite simply, we'd be able to consciously send messages to our past and future selves.

But... what does that even mean?

Is it even possible?

Quite simply, it should be.

Conscious Time

As humans, everything we do unconsciously, we can learn to do consciously.

Sending messages to our future selves isn't what's crazy here. We can already do that. We can write a message on a piece of paper, set a reminder, or even just make sure we remember something important.

What's incomprehensible is the possibility of sending energies to our past selves, messages we hadn't received before.

It's not just crazy... it's nonsensical.

I've been trying to understand the implications of this for nearly a year now. All I've known is that it would compound the power of extuity to a level we can't yet comprehend.

If we could change our past, one would think that we'd change the present. But that's not how time works.

That's because of the "Grandfather Paradox". If we were able to alter the past and stop our grandfather from meeting our grandmother, then we would never be born, which means we couldn't go back in time in the first place, making it impossible to stop them. It's a logical contradiction; a paradox.

Our physical reality cannot be altered by changing our past.

So, people say that if it were possible to alter our past, it would create an alternate reality. In the case of the Grandfather Paradox, it would create an alternate reality we weren't born in, one that exists alongside our reality, thereby not affecting our current reality...

But that doesn't make sense either.

And there's a reason it doesn't make sense, and it's down to an aspect of time we haven't yet covered. In fact, it's one that's quite controversial because it points to one undeniable conclusion:

Alternate realities do not exist.

Let me explain.

Earlier, I presented to you a "Time Map", an intricate model of paths through time where different paths can be taken at "Choice Points". This model is just conceptual. It's not a real portrayal of reality. It falsely proposes that all other paths exist as alternate realities. That model only represents how our path through time works, but it doesn't represent 'time' or reality.

To understand this, it's time we revisited the Quantum Cat.

Schrödinger's Cat.

This representation of superposition – being in all possible energetic states at the same time – is critical to the understanding of our reality. To recap, while the cat is in the box, we cannot observe the cat's state in any way, meaning it is both alive and dead. Its energetic state is in superposition, existing in all potential realities, until we open the box and observe it.

To remind you, this is called the "Observer Effect", or more technically, the "Wave Function Collapse".

This proposed phenomenon was proven in the Double-Slit experiment where particles act as energetic waves until observed, at which point they collapse into their particle form and are forced to pass through one of the two slits, forced to exist in one reality.

What's critical here is that once potential realities are observed, it is *realised*. It becomes *reality*.

There is only *one* realisation, *one* reality.

The one we observe.

At the point that potential energies are realised, all possibilities collapse into the realised reality.

So, at choice points, where we can take one of two paths, we select a reality to step into and observe, meaning the potential energy of that reality is realised. The other paths are not observed. The other paths are not realised. They do *not* exist as alternate realities.

The "Time Map" is a popular representation of time, but it wrongly suggests that the other realities are still there, that there might even be different versions of us in the alternative reality, a different version of us that made a different choice.

It's incorrect.

However... it's not *entirely* incorrect, hence the illusion.

To understand this, it's time to look at the final evolution of the Time Helix.

The Time Helix is, in fact, a fractal.

Yes, a fractal.

If you're unfamiliar with fractals, it's a pattern that is repeated within itself. In other words, if you zoom in on a pattern and find the same pattern, it's a fractal. They're a natural part of nature. Natural fractals include snowflakes, tree branches, and shells.

THE THEORY OF EXTUITY

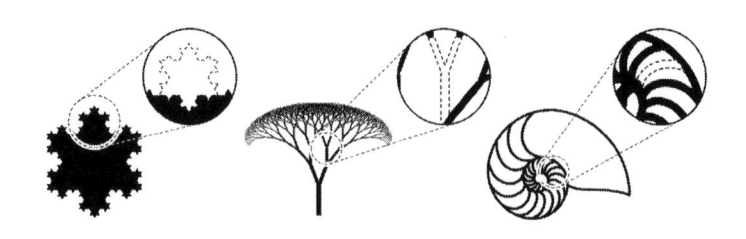

So, how does this work with the Time Helix? Well, if we zoom in on one of the timewaves, because it's a fractal, we'll find another Time Helix.

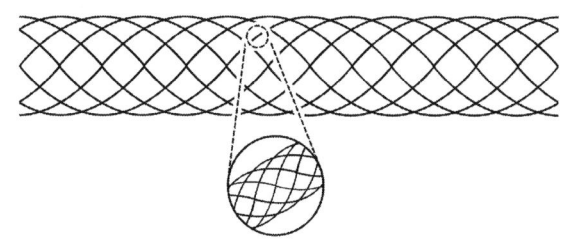

Pretty cool, right?

Now we get to dive into a critical aspect of the Time Helix... how it transforms over time, as it moves forwards into the future. It revolves around the fact that each timewave is another Time Helix, a group of multiple timewaves.

To simplify this, let's look at an example scenario.

You're looking to pick up a new sport to keep active. You research local sport clubs and find a few options; tennis, swimming, rock climbing, dance etc. Now, even though you've never tried any of them before, you feel an odd pull to rock climbing. You feel like you'd really enjoy it.

This, as we've covered, is a Choice Point.

Now, your Time Helix is made up of timewaves that each represent a specific part of your current reality, and one of these timewaves would represent the sport you decide to take up. Let's call it the sports timewave.

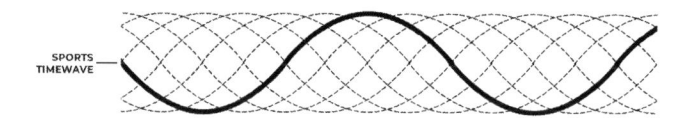

Remember, each timewave is itself a Time Helix, meaning the sports timewave actually contains multiple timewave energies inside it.

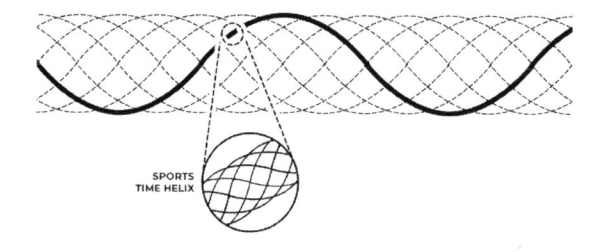

Now, in the scenario of your sports timewave, it would actually contain multiple timewaves for each of the possible sports that you could do. Each timewave represents different energetic potentials for each sport, including rock climbing.

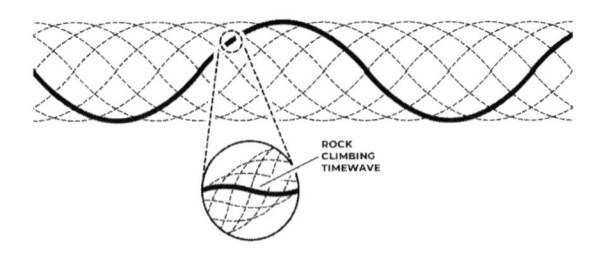

THE THEORY OF EXTUITY

[Figure: Rock Climbing Timewave shown as a thick wave separate from the dashed Time Helix, with a magnified circular inset]

So, what happens *if* you choose a sport? What happens when you make a choice and "pick a path"? Well, the timewave that represents the sport you choose separates from the Time Helix that makes up the sports timewave, and it joins your immediate Time Helix. In effect, the potential sport option now becomes a realised sport activity in your reality.

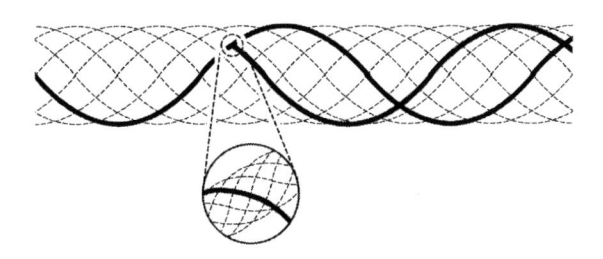

[Figure: The rock climbing timewave now merged with the immediate Time Helix]

It's a "Timewave Collapse".

This is what happens at Choice Points.

At the moment the sports timewave collapses with one of its inner timewaves merging into your reality, there is a Time Helix reaction. You can think of it as an emotional reaction to the new reality you've entered, to the new sport you decided to take; you could absolutely love it, hate it, or find it completely boring. It can even affect every other timewave in your Time Helix. This is the emotional energy that your future self would experience on this 'path', the potential energy that you can sense in the past.

But what happens to the other sport timewaves? What happens to the other energetic potentials of the other choices that you didn't make? Did they simply disappear?

No.

Energy cannot be created or destroyed. It's one of the fundamental laws of physics.

The energy still exists in the inner Time Helix, but it remains as *potential energy*.

The *energies* of the "alternate paths" or the "alternative realities", the realities that were not realised, exist *in* your reality as potential energy. They remain in the field of energy behind our reality, or as it's commonly called, the "Quantum Field".

There is no such thing as "alternative realities", only "energetic potentials" of unrealised realities.

Which brings us back to "Conscious Extuition".

If we could consciously send energies to our past selves, messages we hadn't received before, what actually happens? Altering the past cannot change the present reality as the Grandfather Paradox proves, and if an "alternate reality" cannot be created, what else is there? What would this ability allow us to do?

Nearly a year. I spent nearly a year pondering this.

Eventually, I discovered what it meant. I discovered what the theory proposed would happen.

Our past selves are, quite obviously, our 'past' selves. We had specific beliefs, opinions, and values. Our state of mind was what it was. We played out our time, made the choices we did, and lived out the days and years we did. We were so ingrained with what made us who we were, we ended up where we are today.

As we covered earlier, one of the most powerful ways of redirecting our future paths is by *creating* choice points. The way we do that is by logically questioning our deepest beliefs and our deepest opinions, everything that makes us who we are. We create questions where there would not have otherwise been one, hence creating new choices we can make day to day, creating entirely new potential futures.

That's the logical route to creating choice points.

And there's another.

Conscious Extuity.

If we were able to be *consciously* extuitive, to consciously send out emotional energies across time, we effectively have the ability to create a strong feeling of intuition in our past selves, a *new* feeling of intuition that we didn't feel before.

That means we could, in the present, create choice points for our past selves.

I repeat, we could *create choice points for our past selves.*

Through conscious extuition, if we could force the energy of our past selves to 'feel' differently about a specific opinion, one we've never questioned before, we would force our energetic past selves to experience a new choice point.

And what would a new choice point create?

It would create a "new path", or as we now know, a new "energetic potential" of an alternative reality. It would create a new energetic past. In other words, conscious extuition would change the energy of the past.

Which, by the way, would have a profound effect on an aspect of humanity.

One that few realise is also deeply intertwined with emotions.

Memory.

Psychologists know this all too well.

In fact, they have a common saying: "You'll forget what people say, you'll forget what people do, but you'll never forget how they made you feel."

And it's true.

Our strongest memories are moments of strong emotion. We'll never forget those moments that made us feel truly happy or severely depressed.

Memory has a deep connection to our emotions.

And I also believe that memory has a deep connection with the Theory of Extuity, or specifically, the Time Helix. I believe that this new model of time reveals the true explanation for one of the greatest phenomena we experience with 'time', one that directly relates to our memory.

You've heard of it before.

Deja Vu.

It's one of the greatest mysteries of mankind.

When one experiences Deja Vu, they have the inexplicable feeling that they lived that moment before. They 'remember' being in that

THE THEORY OF EXTUITY

situation before, but know they haven't. Their memory is playing tricks on them.

Or is it?

Deja Vu was one of the most fascinating mysteries that I tried to find an answer to very early on, and I found one that made sense to me.

Imagine this...

Imagine you're outside and there's a ball on the ground. There's no one around. It's just you in an open space. Now, you're deciding to do one of two things: Kick the ball so it rolls along the ground, or slightly chip it so it bounces.

In other words, you're at a choice point, and there are two potential futures ahead of you.

But this time, these aren't two separate realities.

Imagine that in this scenario, whether you kick the ball so it rolls or bounces, both choices would end up with the ball stopping in exactly the same position at the same time. What happens in terms of 'time', or the energy of the reality, is that the choice point would create two "paths" through time, but those two paths would actually merge back into one identical state of reality.

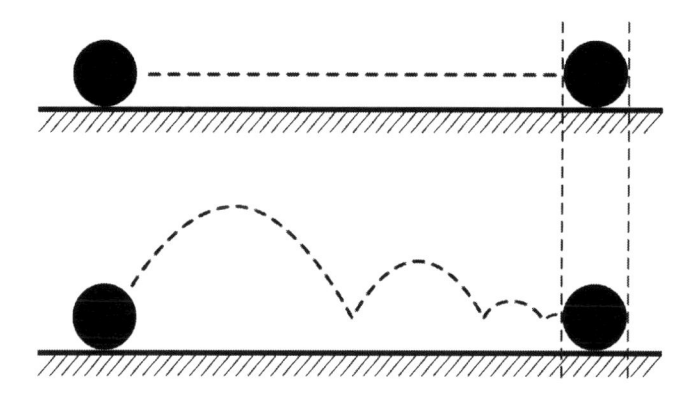

The potential energy of both realities would split, and then merge back together into one reality simply because the energy of both realities would be absolutely identical again.

This is how it would look in the traditional "Time Map" model:

And how would it look in a Time Helix? We'd see a timewave collapse when we choose to kick the ball so it rolls or bounces, and then when it stops, we'd see the unrealised energy potential of the other reality, the other timewave we didn't choose, merge with the timewave that we did observe, simply because it would be an identical energy.

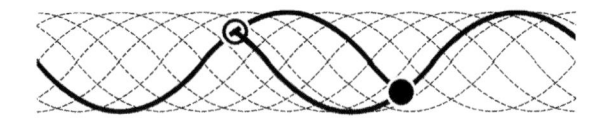

What happens here is that the energy potential of the unrealised alternate reality has reached a state that is *identical* to our current reality, and therefore merges into one. Despite the alternate reality not having been observed, the identical *resultant* energy has fused it with our physical reality.

I call it a "Fusion Point".

It would be an *extremely* rare occurrence in our reality with so many variables changing all the time, but it's not about the collective reality, it's about our individual realities, our individual flow through time. It will still be rare, but when it does occur, the result is quite spectacular.

At a fusion point, we have multiple potential pasts.

It's the *opposite* of a choice point, where we have multiple potential futures. Choice points create potential futures, while fusion points create potential pasts.

And just like how a choice point works, we can *feel* emotions and energies from both potential timewaves.

As you can guess, this is what I believe causes Deja Vu.

Deja Vu is also an *extremely* rare occurrence. It's a state where we strongly feel like we've been in the same situation before, but know we haven't. It's a conflict of emotion vs logic. It's a conflict of energy vs reality.

I believe that those moments we 'feel' we've been in before also occurred in the alternate potential reality which merged with our own. It didn't happen in the past of *our* reality, but it did happen in the potential reality whose energies ended up becoming identical to our own.

It occurred in the new potential past.

And there's something critical to remember here: A fusion point, no matter how impossible it may be, can *only* occur after a choice point. A choice point creates a split, forms energetic potentials of alternate realities, and fusion points merge them back into one.

And that is what makes the potential of conscious extuity so powerful.

More powerful than you can imagine.

If, through conscious extuition, we consciously send out strong emotive energies that make our energetic past selves question an opinion, belief or pre-conditioned fact in our mind that we would never otherwise have questioned, we would create a choice point.

By creating a choice point in our past, we'd create a new potential alternate reality.

By creating a choice point, we'd potentially create new fusion points.

By creating new fusion points, we'd create new potential pasts, and with it, new memories.

Yes, one of these memories may be an event that we'd also experience in this reality, hence Deja Vu, but there would also be ones that would be unique to the alternate past.

The "false" memories.

Of course, no one would believe anyone who experiences it. There would be no physical evidence of these memories because the energies wouldn't occur in the physical reality. But, if this were possible through conscious ability, surely this would also occur by accident too?

Surely, "false" memories would be another phenomenon of humanity?

Well, guess what... it is!

This phenomenon *does* exist in our reality, and it works *exactly* as this theory predicts it would.

And, oh my, is *this* a mind-bending one.

It's time for one final story, one from recent history, about history...

It's 2009.

Author and researcher Fiona Broome is at a convention. During the event, she overhears a member of the security team casually mentioning how people "remembered" when Nelson Mandela died in prison in the 1980s.

But Nelson Mandela hadn't died in prison. In 2009, he was still alive.

Broome *froze*.

She also "remembered", very clearly, that Nelson Mandela died in prison, of news clips of his funeral, the mourning in South Africa, some rioting in cities, the heartfelt speech by his widow. That was until a few years beforehand when she discovered, to her surprise, that Nelson Mandela was alive. At the time she shrugged off the thought, thinking she may have misunderstood something on the news.

LEVEL 7

Except, she just heard that someone else had this exact same "memory".

She *had* to investigate.

In the months that followed, Fiona Broome created a website to see if any others reported having the same memory. To her surprise, a large community of people responded, strangers who remembered the *exact same* Mandela history.

A history that *never* happened.

Nelson Mandela was a famous South African anti-apartheid activist who served as the first president of South Africa from 1994 to 1999. As you may recall, he died in 2013.

2013, as it turns out, surprised quite a few other people too.

When news coverage of his *actual* death circulated, many people reacted in the same way... "I thought he died in the 1980s."

This is a fact. You can find numerous reports across the web.

Reddit. Blogs. YouTube.

People around the world, complete strangers to each other, all have vividly detailed memories of news coverage of Nelson Mandela dying in prison in the 1980s, of his funeral, the mourning in South Africa, some rioting in cities, the heartfelt speech by his widow...

Complete strangers across the world have a collective "false" memory.

Fiona Broome dubbed this phenomenon the "Mandela Effect", and it turns out it wasn't the only shared "false" memory that people reported.

Maybe you can relate to one of these other false memories.

Many report the logo of clothing brand "Fruit of the Loom" featuring a cornucopia, but it doesn't, and never has. The company themselves confirmed they didn't.

Others report Mr. Monopoly from the board game wearing a monocle, but he doesn't and never has.

Many more report – including salesmen at video rental stores – a 1990s movie titled "Shazaam" starring comedian Sinbad as a genie, yet both Disney and Sinbad confirm it never existed. Yet, video store owners distinctly remember ordering and receiving copies of the movie and putting up posters in their stores.

Yet, none can be proved.

None can be proved because there is absolutely no evidence of any of it.

None, at all.

There is only "memory", the same memory from countless strangers.

Many claim to have 'debunked' some of these based on psychological influences from other sources, and it's important to understand that there *will* always be an element of psychology around every experience with time, emotions and memory, and some of these may indeed be purely psychological. However, there is never just one answer, and there has never been an answer for all of these false memories.

Until now.

These false memories are exactly what "Conscious Extuity" proposes would happen. Or, specifically, what "Fusion Points" in time would cause.

To help explain this, I'd like to share a simplified example.

Imagine you're heading to a park to read a book and you decide to walk instead of taking a bus. You arrive at 11.10am, and in the alternate potential reality on the bus, you would have arrived 10 minutes earlier at 11am. Upon sitting on a bench, the energies of both realities become identical because you'd also be sitting at that exact same bench in the alternate potential reality.

So, both realities merge into one.

It's a fusion point.

Now, immediately after taking a seat, you get sense of Deja Vu. You feel you've been there before, at the same bench, in the same weather, on the same day. Why? Because the fusion point added a new potential past, meaning you now have *two* pasts. That brought with it the "false" memory of being on a bus, and the exact same memory of sitting at that exact same bench on the exact same day, which occurred 10 minutes earlier in the alternate past.

Once again, it's extremely rare for alternate paths through time to result in completely identical states of reality – i.e. in this example, arriving 10 minutes earlier via a bus can't have affected the environment differently in *any* way, such as startling birds or stopping another from sitting on the bench – which fits because Deja Vu and the Mandela Effect are themselves also rare.

At a fusion point, the energy potential of the unrealised alternate reality fuses with our current reality.

At a fusion point, we have multiple potential pasts.

At a fusion point, multiple pasts mean multiple memories. "False" memories.

The memory we'd "remember" would be the one that is most emotive to us, the one that triggered the greatest emotions.

That's how memory works.

There would be no physical evidence of the memories from the alternate energetic reality that fused with ours because the energies weren't observed in our physical reality. No observation means no physical materialisation.

Just multiple pasts. One physical past, but multiple potential pasts.

But, let's say this could be done consciously, intentionally, what would be the point? I wasn't sure until I read through the first-hand reports of people who experienced the Mandela Effect.

When I did, something really interesting struck me.

These "false" memories weren't *just* memories they could recall in detail. Individuals also report having *learnt* something when they

LEVEL 7

"experienced" these memories, *knowledge* they wouldn't otherwise have known.

Knowledge they didn't pick up from their real past.

When they "remember" Nelson Mandela dying in the 1980s, many remember asking about "apartheid", the racial segregation that Mandela fought against. They recall that as the point they learnt what it meant.

When they "remember" seeing Fruit of the Loom's logo with a cornucopia, many remember asking their parents what the object behind the fruit was. They recall that as the point they learnt what a cornucopia was. Yet there never was a cornucopia in the logo.

Which is crazy.

And in this instance, I mean crazy powerful.

If it's possible to consciously create new potential pasts, it effectively means we could "learn" things without ever having experienced them.

New knowledge. New experience. New skills.

It's crazy. Ridiculous. Incomprehensible.

Conscious Extuity.

It reminds me of the scene in the original Matrix movie where they connect Neo to a computer and "download" skills to his mind.

If it were a real possibility, it would be unbelievably powerful.

Conscious extuition creating choice points.

Choice points creating fusion points.

Fusion points creating new pasts.

New pasts creating new memories.

New memories creating new skills and experience.

THE THEORY OF EXTUITY

Possibly the most powerful ability of humanity.

Conscious Extuity.

And remember, this is just *one* application of Conscious Extuity, the only one that I've managed to discover so far...

So, where on earth do we go from here?

Becoming Timeless

I thought of a great analogy.

Time to humans is like water to fish.

We live in it, breathe it, are completely subject to its power.

Yet, we can navigate through it. We can go quicker or slower. Go in any direction we like.

A wave may come, pure energy, and it will sweep us in a direction not of our choosing, but nothing's stopping us from finding different currents, riding the flow of energy, and rewriting our journey onwards.

And while past waves of energy push us in, future tides of energy pull us out.

Tides may pull us towards a place we'd like to go, or a place we wish to avoid. But, by *feeling* them, and consciously being aware of them early enough, we can *choose* to go with the flow or avoid it altogether.

Time is an ocean of waves. Timewaves. Pure energy.

You can just keep swimming, or strive to understand its power and use it to propel you towards a place of your choosing.

And the source of that power? The foundation of that potential?

The Theory of Extuity.

"All emotions radiate outwards from the point in time it is experienced and can be felt at any other point in time, with an energy proportional to the strength of the emotion and inversely proportional to the distance of time between them."

Extuition is an ability all humans possess, and it's an ability that makes mastering time possible.

And as I mentioned earlier, the Theory of Extuity isn't just the final definition. Extuity includes everything that makes extuition possible. Extuity includes a true understanding of time.

So, how about one final recap of our journey...

On Level 1, we uncover the flow of time is the flow of energy, and more specifically, time is energy. As we have experienced with our GPS satellites, any flow of time can be sped up or slowed down, a phenomenon called "Time Dilation". It was just as Einstein proposed in his Theory of Relativity. The greater the energy an object experiences, the slower its flow of time, and the faster time moves around it. Time is therefore not a universal constant, it's different for every object in the universe. Furthermore, each unique flow of time is not constant either; its rate of flow is ever-changing. Finally, we introduce "timewaves", a new model of a timeline that represents time as energy.

On Level 2, we dive into Quantum Physics, "Quantum" literally meaning 'energy'. On this level, we discover how the "Double-Slit Experiment" reveals energies can be in all possible energy states at the same time until observed; "Quantum Superposition". Applying this phenomenon to the "Quantum Cat", we evolve the timewave model into a "Time Helix", representing how one entity can be on multiple timewaves at the same time. With superposition, we discover how energies can be synchronised or 'entangled' across space, known as "Quantum Entanglement". More profoundly, it also proves energies can be synchronised across time with a clear stance on it being directionless; the energies could be sent both forwards and backwards through time.

LEVEL 7

On Level 3, we investigate the connection between emotions and our bioelectricity. Understanding emotions are labels for our energetic states, that emotions are energies, we explore how we experience time in differing states of emotion. Outlining how states of high emotion make time seem to fly by and how states of low emotion make time seem to slow down, we unveil a startling connection to Time Dilation which we call "Biological Time Dilation". We highlight the connection that emotions have with time and explored the "Energetic Body" counterpart to our physical human body, our "Quantum Body".

On Level 4, we dive into another phenomenon of the human experience. We uncover how "intuition" – an accepted mystery of humanity – is an experience that directly involves emotions. It specifically occurs when emotions transcend space or time. There's a clear connection between the characteristics of intuition and Quantum Entanglement, leading to the proposal of "Biological Quantum Entanglement". Our focus, the time aspect of it, was validated by a research study by the HeartMath Institute. Before diving into the implications of this, we are taken to Level 5.

On Level 5, we uncover that intuition – the ability to receive emotional energies across time – has a counterpart. By definition, anything that is 'received' must be 'sent'. This means emotional energies are sent across time, the opposite of intuition. There is no word for it, but exploring a possible 'opposite' leads to the word "extuition". As "intuity" is the word defining one's ability to intuit, or be intuitive, "extuity" is the word defining one's ability to extuit, or be extuitive. We begin exploring the implications of the theory by looking at how extuitive energies from our past affect our present, analysing 'grief'.

On Level 6, we continue on to look at the more incomprehensible side of the theory: how extuitive energies from our future affect our present. By analysing the "language" of extuity, our emotions, we decode "coincidence" and "love at first sight", discovering how opinions are influenced by extuition and how our lives from childhood are influenced by our futures. Uncovering how our consciousness amplifies our energetic connections through time, we introduce "Choice Points" on our "Time Maps", points in time where we can choose from one of many potential future paths. We discuss how conscious focus can help direct us at choice points to futures where we experience our greatest energetic potential, and the correct questions to ask in order to use this "Time Compass". We also highlight how our ego can hinder us, and how lost ancient teachings have left us so vulnerable to that trap. We further discover how logic can be used to create choice points and entirely new futures. And, like all compasses, a destination is needed, so a vision of our ideal future is key. We round off by answering whether 'fate' exists.

On Level 7, we dive into humanity's greatest source of power: becoming conscious of our unconscious abilities. After visiting Conscious Breathing, Conscious Bioelectricity and Conscious Dreaming, we're introduced to "Conscious Extuity". The power of this is unclear, so we dive deeper into the dimensions of time. Upon learning about the illusion of Time Maps and the fractal nature of Time Helices, we rule out the creation of "alternate realities". We

LEVEL 7

instead dive into "Fusion Points" and how they give rise to experiences that we describe as "Deja Vu" and the "Mandela Effect". As choice points create potential futures, fusion points create potential pasts. Alongside this leading to "false" memories, we discover it brings with it knowledge, concluding that conscious extuition could give us the power to "learn" things without ever having experienced them in this reality.

And now, we're here.

We're at the end of all 7 levels of understanding of time.

And, I have a gift for you.

What's funny about this gift is that if you'd like to receive it, your past self has already accepted it.

One last time, let me explain...

You have *consciously* read this book. Your mind has been made aware of new points of view of a deceptive reality you've been hardwired to see. Or rather, your mind has been exposed to entirely new realities that you may never have otherwise considered.

By consciously taking it all in, you've unlocked entirely new pathways in your subconscious mind.

And that's where the true power lies: Your subconscious mind.

Now, don't get this confused with your unconscious mind. Our unconscious and subconscious minds are very different. Our unconscious mind deals with automatic processes that require no awareness or conscious control, e.g. breathing, with 'un-' meaning "not" or "opposite of". Our subconscious mind is the foundation of our conscious mind, 'sub-' meaning "underneath" or "lower". Our subconscious mind influences *how* we think, defining everything we believe about ourselves and the world around us; capabilities, limitations, possibilities etc.

I like to think of our conscious mind as a "programmer", our subconscious mind as our "program", and our unconscious as a "background process". Our memory is, of course, our "database".

So, as you've read this book, you've alerted your subconscious to entirely new possibilities, new understandings of 'time', and new potential abilities. Now, being subconsciously aware of it all, your subconscious mind will encourage your conscious mind to see reality in a new, more powerful way, and you will slowly begin to master time on a conscious level.

Well, that's only if you truly, on a deep subconscious level, wish to unlock a greater potential.

If you do, you have a very interesting time ahead.

You will slowly begin to distinguish "Earth Time" from your own flow through time. Hours and minutes won't define your effectiveness. Days and weeks won't make you stress. "Time flying by" will eventually stop surprising you as you become more self-aware of your personal flow of energy. Your age as defined by calendars will become meaningless as you begin to truly appreciate your input and output of energy in life, everything you embrace and share.

In effect, you'll embrace the true meaning of time day-to-day, just as the Greeks identified. Not "Kronos" time, not chronological, quantifiable time, but "Kairos" time, experiential, qualitative time. The true life embodiment of quality over quantity, the true measure of your life.

And, while it will take longer, you *will* slowly begin to accept, embrace, and utilise your power over time. Your connection to your multiple potential futures will strengthen. You'll navigate difficult Choice Points with greater confidence. You'll experience coincidence, love at first sight, and extuitive powers with deeper appreciation. Your awareness of multiple potential pasts will become more apparent. Moments of Deja Vu will pass by with a faint smile of understanding and appreciation of the raw power of our reality.

You *will* slowly, surely, subconsciously transcend the illusions of quantitative time, embracing the powers of energetic time.

And that, my dear friend, is my gift to you…

A reality beyond possibility…

A freedom beyond freedom…

Becoming timeless.

Side-Level 7

A Final Gift

———

Memory.

It's a funny thing.

It's completely intertwined with time, yet, it's one-directional. It exists only for past events.

Or does it?

Before we consider that, I'd like to share a phenomenon that reveals an aspect of memory that few have heard of.

"Heart Memory".

Paul Pearsall, MD, a psycho-neuro-immunologist, discovered a shocking phenomenon after interviewing nearly 150 heart transplant recipients.

Hearts carry memory.

There's one particular story reported by Pearsall that highlights this phenomenon perfectly, the story of an eight-year-old girl who received the heart of a ten-year-old girl who had been murdered.

Upon receiving the heart via a transplant, the recipient had horrifying nightmares of being murdered by a stranger. Psychiatric help was sought because of how traumatic they were, and the images the girl saw in the "dream" were so specific that the psychiatrist and the mother notified the police.

THE THEORY OF EXTUITY

According to the psychiatrist, "...using the description from the little girl, they found the murderer. He was easily convicted with the evidence the patient provided. The time, weapon, place, clothes he wore, what the little girl he killed had said to him... everything the little heart transplant recipient had reported was completely accurate."

The recipient of the donor didn't just receive a heart, she also received memories from the donor.

This is one of countless stories that conclude the same reality: memory is not just a function of our brains. Somehow, our hearts play a role too, which reminded me of a particular research study that we covered in Level 5.

The HeartMath Institute study that proved intuition across time.

If you recall, they discovered that the "intuitive signal" originated in the participants' hearts before being passed to their brains. The energy came from the heart. Taking this into account alongside Pearsall's research, it cannot be denied that our hearts play a role in connecting us to our emotions across time and to our memories from our past.

Interesting.

It seems that "memory" may be hiding a secret.

As we already know about memory, it has a deep connection to our emotions. The stronger an emotion, the stronger the memory. And as I mentioned in Level 6, I believe that memory has a deep connection with the Theory of Extuity, which leads me to a final gift I wish to share.

As we know, the beliefs of our possibilities and limitations that exist in our subconscious minds define the very possibilities and limitations we experience in our conscious life, and there's one specific 'fact' that we've been hardwired to believe since childhood...

A memory is something remembered from the 'past'.

Just the past.

But what if we truly believed, on a deep subconscious level, that memory was omnidirectional? That memory applied to both past and future? Would we "remember" moments in our future, specifically those in which we experience strong emotions, exactly like we remember our past?

In other words...

Would we have "memories" of our future?

I'm going to let you discover the answer to that one.

Have fun.

YOUR JOURNEY
Level Progress & Key Learning Points

1. Time is energy
+ Time is elastic + Time is not constant

2. All points in time can be connected
+ One can be on multiple timelines

3. Emotions are energies
+ Biological time is elastic + Quantum bodies exist

4. Emotions transcend time
+ Intuition is receiving energies across time

5. Extuity influences past & future
+ Extuition is sending energies across time

6. One can write futures
+ Choice points create multiple futures

7. One can rewrite pasts
+ Fusion points create multiple pasts

Final Words

The Journey's End

Albert Einstein said, "If at first an idea does not sound absurd, then there is no hope for it."

So, let's lay out just how beautifully absurd everything we've covered is...

Time is energy. It's unique to every entity. It can be sped up and slowed down. All points in time can be connected. Emotions transcend time. We are connected to all past and future selves. The future, as well as our past, is writing our present. We can feel our futures and decode them. We can create and navigate multiple potential futures. We could potentially create and merge with multiple potential pasts. Natural human phenomena could be controlled. Egos can be mastered, and fates can be rewritten. We can transcend the illusions of time, becoming timeless.

The Theory of Extuity.

Beautifully absurd.

Maybe there is hope for it.

However, there is one thing I'd like to point out that is extremely important for everyone to know...

Just because I've stumbled across a temporal element of humanity, it doesn't mean I've discovered every possible implication of it. To put that into perspective, just because one discovers ink, doesn't mean they've discovered every possible use of it. Stories, poetry, maps, art and more are all a result of the collective human race.

Similarly, extuity is like a form of ink for time. It enables the ability to both read and write time. Its implications and applications are *limitless*.

To emphasise once again, this book is just the beginning.

There's *so* much more to uncover here, both in how extuity affects us and how we can use it. It's the foundation of our connection to our past and future. It reshapes our *entire* reality.

So, think about it, discuss it, share your ideas.

In fact, I'd actually recommend thinking of someone you know – just one person – who will find this topic as captivating as you do. Someone who would absolutely love to discuss the limitless possibilities of this theory. Someone you could even test some of the applications of this theory with.

Let your mind wander freely with them.

And remember, experience is and always will be the greatest teacher. I'd highly encourage you to flick through this book a few times, highlighting key aspects of extuity that you believe you can work on, and put it into practice. I mentioned at the end of Level 7 that your subconscious mind will naturally embrace a more timeless nature, but you can consciously embrace it faster and more intentionally.

And that's where I'd like to share three 'powers' to help you do just that...

A logical power.
An emotive power.
And, an ancient power.

Timeless Powers

I'm not one to just discuss cool topics, I'm one who likes to discuss topics that can be applied in effective ways. It makes the time spent talking more meaningful. Naturally, the Theory of Extuity ticks that box as it will allow one to navigate their many potential futures to live a life they've always dreamt of, but there are few specific areas where further advice can be helpful.

Which brings us to the first of three powerful tips, a logical power.

Whilst we dived into navigating choice points in time with our "Time Compass", there's one aspect of time we didn't cover: how to use the time in between choice points.

I'd like to briefly touch on this as it is something I'm known for; time management on an "impossible" level.

I'm able to build entire apps in days that would take companies months. I'm working on multiple tech companies whilst finding time to hike, rock climb, visit family, and travel. I utilise my time with unprecedented efficiency because of a very unique time management technique that I designed many years ago.

I call it a "Priority List".

It combines and simplifies the power of Eisenhower, Priority Matrices and Kanban principles.

My algorithm behind it is a tad bit complicated so I'm not going to dive into how it works here. I have something better for you. I originally utilised this technique by building a model in Excel spreadsheets, but it was an overly complicated interface. So I decided to build a *very* simple app for it.

FINAL WORDS

You can find it at: www.prioritylist.app

Priority List
Scan QR to visit: **www.prioritylist.app**

If you are someone who feels like they have endless tasks to do, it'll be extremely powerful for you. You will essentially master your flow of time in between the big choice points in your life. Working smarter, not harder. Using your time more wisely. *Efficient* application of your energy.

It's worth trying.

That's the first of the three powerful tips I wish to share.

The second, the emotive power, is one I've mentioned many times in this book: Mastering Emotions.

More specifically, *how* to master your emotions and ego. Just because we've lost our ancient teachings, doesn't mean we can't rediscover those teachings.

Thankfully, society today has done the hard part. With the great focus on Mental Health and Mental Fitness, we've discovered nearly everything we need in order to master our emotions. The only downside is that these teachings are fragmented, meaning there isn't "one place" to find everything we need regardless of the state of our mental fitness.

Well, that was until I started working on solving that problem in 2023.

Emotional intelligence is key to mastering our flow through time *because* it's key to mastering our very being. I didn't just want to achieve it personally, I also wanted to help loved ones achieve it too, so, naturally, I built a few apps.

Remember, I'm known for building beautifully simple, easy-to-use apps, and I wanted to apply that skill to mental health.

My favourite is one I built in 2024. In just its first two months it had over 40,000 users with endless praise of how it had helped individuals regain control over their emotions *during* extremely intense moments, including panic and anxiety attacks.

I've wrapped these apps in a brand called "Heartling", a name representing the apps being seedlings of heartfelt growth.

You can find them here: www.heartling.org

Heartling
Scan QR to visit: **www.heartling.org**

Check it out, master your emotions and your ego, and begin mastering the use of your Time Compass to master your flow through time. It's key to transcending the false illusion of quantitative time. It's key to becoming timeless.

Now we come to the third power, an ancient power.

It concerns the greatest potential of extuition, Conscious Extuity.

FINAL WORDS

As you may have realised, I didn't state *how* this can be done, *how* one can consciously extuit emotional energies. It's because I don't have an answer.

Or, at least an answer that I'm confident in just yet.

But I do have an idea of how we can unlock Conscious Extuity, and it's with an ability that I've mentioned many times in this book: Conscious Bioelectricity.

As I wrote in the introduction, after I broke my arm while snowboarding in 2023, my journey into the mind led to the ability of *"feeling and directing the flow of Chi in my body"*. I studied many forms of meditation while waiting for my arm to heal, and after months of practice, I began to feel this unfamiliar sensation that I could move around my body at will.

At first, I had *no* idea what it was.

Unlike emotions, our lost ancient teachings have made it nearly impossible to rediscover teachings around our bioelectricity and our body's electromagnetic field. It's not taught *anywhere* in our modern society.

But that didn't stop me.

I figured there was *some* literature, *some* teachings passed down from ancient cultures, and I went on a hunt to unearth them. Eventually, I found some within Ancient Chinese Medicine that call this aspect of our human nature "Chi", hence why I also call it Chi.

Through studying Chi, I've been able to strengthen my control over what I discovered to be my bioelectricity, and I've found many applications for it that I've mentioned throughout this book.

What's fascinating about it in relation to the Theory of Extuity is that I've managed to use it to directly reprogram my emotions. I've simply focused on thoughts that trigger low emotions while inducing a bioelectrical state in my body of deep peace. It manages to "overwrite" the emotion that the thought triggers.

Conscious Bioelectricity *has* to be the key to mastering Conscious Extuity, because emotions *are* the language of extuity.

Unfortunately, I've yet to come across literature that actually teaches how one can *become* conscious of their inner energy. All I've found so far surrounds cultivation and application, but not *identification* of it. Everyone I've met who shares this ability seems to have also become aware of it by chance or by pure instinct. It seems this specific teaching remains lost to ancient times.

So, I decided to write the handbook myself.

I began by trying to help a handful of friends and family become conscious of their Chi – their bioelectricity – with some techniques that I came up with myself. It took a while as many of the techniques didn't work, but eventually I found some methods that allow one to become conscious of their Chi in as little as five minutes.

At this point in time, Autumn 2024, I haven't published any of it yet.

Rest assured, I plan to.

I've named the project "Chi Fu".

Identification, cultivation and application of Chi, of bioelectricity.

You can find it at: www.chifu.org

FINAL WORDS

> **Chi Fu**
> Scan QR to visit: **www.chifu.org**

If you're wondering about the name, I was inspired by "Kung Fu". "Kung" means 'work', while "Fu" means 'time spent on'. Similar to Kung Fu, "Chi Kung" (or "Qi Gong" in Hanyu Pinyin) is a practice that involves physical exercise for health, like a 'slow' yet very controlled version of Kung Fu. However, Chi Kung focuses on physical movements to cultivate and encourage the flow of Chi, whereas my proposition of "Chi Fu" is to rebirth the ancient primary focus on Chi, becoming totally conscious of it as to not require any physical movement to cultivate it and control its flow.

Chi Fu.

Ancient teachings, reawakened.

The ancient key to mastering time.

Quantum Shifts

There's one thing that extuity makes perfectly clear...

We are not just energy, we are *conscious* energy.

There are countless abilities waiting to be discovered, and it all starts with *understanding* our energetic elements. It's not something easily done alone, but I believe this book is the perfect catalyst to change that.

This book hasn't just introduced a revolutionary discovery of time, it's a symbol of life-altering understanding and growth. 'Extuity' is more than just a theory; it represents open-minded individuals with a deep interest in theories and abilities that we've yet to discover and master.

The "Extuity Community".

A global community of open-minded souls living to explore the unexplored.

Whether it's diving deeper into time, exploring lucid dreaming, chi, or even out-of-body experiences, nothing's too 'impossible' for us to discuss and share experiences of. We believe anything's possible. We accept that we don't know what we don't know. We will inevitably have greater theories about life, the universe, and everything within.

One of the values that led me to discovering extuity, and a value I wish for us to build this community with, is beautifully captured in one of my favourite quotes, a quote by Steve Jobs...

"Here's to the crazy ones. The misfits. The rebels. The troublemakers. The round pegs in the square holes. The ones who see things differently. They're not fond of rules. And they have no respect

FINAL WORDS

for the status quo. You can quote them, disagree with them, glorify or vilify them. About the only thing you can't do is ignore them. Because they change things. They push the human race forward. And while some may see them as the crazy ones, we see genius. Because the people who are crazy enough to think they can change the world, are the ones who do."

I absolutely love it.

If you resonate, I invite you to join and help grow this beautiful open-minded community. Share your own seemingly-impossible experiences, hear about breakthroughs in any and all energetic abilities, and be the first to find out if I release a future book on the Theory of Extuity.

Join at: extuity.com/munity

Extuity Community
Scan QR to visit: **extuity.com/munity**

Oh, and on the topic of a second book, I'm still unsure of that. I'm certain I'll discover more around the topic of time and our relation to it, and I know there'll be some impactful contributions from the wider community, but whether it'll be enough for an entire 2nd book, I'm unsure. A part of me feels like I will at least release a 'Level 8' at some point, because there *is* something I didn't feel I was meant to include in this book…

The "collective" side of time.

THE THEORY OF EXTUITY

That includes the "Collective Observer Effect", the combined Observer Effect across all living beings, and something I call the "Time Field", a field containing every being's Time Helix and how they interact with each other.

The seven levels we've just covered address you as an individual, but it doesn't dive into how your flow through time affects another's, and how another's affects yours. For example, how do intertwined paths affect navigation through time? Is 'karma' real, and if so, how? And why is it that the Mandela Effect only affects a certain percentage of humanity, and not everyone?

The answers lie in the Time Field and the Collective Observer Effect.

Hence, a potential future Level 8... and maybe 9 and 10 too.

Let's see.

The Final Words

One of the greatest things I've learnt from my business ventures is that no one achieves anything great alone. So, with that in mind, I'd love to ask you for a very small favour to help me build a world where more of humanity embraces impossibilities with deeper appreciation of our natural powers and abilities.

A book review.

It will mean the world to me if you could leave a simple review for this book on Amazon.

Reviews have a bigger impact than you know, especially for unique books such as this. It will help everyone who finds themselves on the fence about reading this book, wondering if it's worth their time to discover the Theory of Extuity.

So, if you can, please take a minute and leave a quick review.

Amazon Review
Scan QR to visit: **extuity.com/review**

Just a heads up, even if you didn't purchase this book, you can still leave a review; simply scroll to the bottom of the Amazon product page and hit "Write a review". If you did purchase this book, you'll see "Add a review" at the top of the product page or in your order history.

You have my deepest gratitude.

And with that, it's time to bring this book to a close...

Henry Ford once said that if he asked what people wanted, they would have said faster horses. Henry Ford invented the car. It's a beautiful reminder that if you want to experience your greatest potential, to accelerate *beyond* your current reality, you must be prepared to *let go* of it.

There's absolutely nothing wrong with sticking with faster horses, but if you're still with me, let's continue this journey together...

A journey of wondrous possibilities...

A journey into a *new* reality...

With love,

Hakeem xx

FINAL WORDS

Think of one person who will find this as captivating as you do.

One who would absolutely love to discuss this theory.

One you can let your mind wander freely with.

Share The Book

Scan QR to visit: **extuity.com/share**

NOTES

References, Research, Studies, and more...

To easily access all links, simply visit: extuity.com/notes

Book Notes
Scan QR to visit: **extuity.com/notes**

INTRODUCTION

1; WEBSITE

Hakeem Javaid

"Designer, Developer, Entrepreneur"

hakJav Studios

https://www.hakjav.com

LEVEL 1

1; WEBSITE

"Early Clocks"

National Institute of Standards and Technology

https://www.nist.gov/pml/time-and-frequency-division/popular-links/walk-through-time/walk-through-time-early-clocks

2; WEBSITE

"Ancient Calendars"

National Institute of Standards and Technology

https://www.nist.gov/pml/time-and-frequency-division/popular-links/walk-through-time/walk-through-time-ancient-calendars

3; ARTICLE

Tolga İldun

"Discovering a New Neolithic World"

Archaeological Institute of America

(2024)

https://archaeology.org/issues/march-april-2024/features/discovering-a-new-neolithic-world/

4; ARTICLE

Peter van den Hoek

"The Solar and Lunar Calendar of Karahan Tepe"

Hunebed Nieuwscafé

(2024)

https://www.hunebednieuwscafe.nl/2024/04/the-solar-and-lunar-calendar-of-karahan-tepe/

5; SCIENTIFIC PAPER

Albert Einstein

"On the electrodynamics of moving bodies"

(1905; vol 2, doc 23)

https://einsteinpapers.press.princeton.edu/vol2-trans/154

6; SCIENTIFIC PAPER

Albert Einstein

"Does the inertia of a body depend upon its energy content"

(1905; vol 2, doc 24)

https://einsteinpapers.press.princeton.edu/vol2-trans/186

7; SCIENTIFIC PAPER

Albert Einstein

"On the inertia of energy required by the relativity principle"

(1907; vol 2, doc 45)

https://einsteinpapers.press.princeton.edu/vol2-trans/252

8; SCIENTIFIC PAPER

Albert Einstein; Marcel Grossmann

"Covariance Properties of the Field Equations of the Theory of Gravitation Based on the General Theory of Relativity"

(1914; vol 6, doc 2)

https://einsteinpapers.press.princeton.edu/vol6-trans/18

9; DICTIONARY

"Kronos (Χρόνος)"

LSJ, Perseus Digital Library

https://www.perseus.tufts.edu/hopper/morph?l=xro%2Fnos&la=greek&can=xro%2Fnos0

10; DICTIONARY

"Kairos (Καιρός)"

LSJ, Perseus Digital Library

https://www.perseus.tufts.edu/hopper/morph?l=kairo%2Fs&la=greek&can=kairo%2Fs0

11; JOURNAL ARTICLE

John E. Smith

"Time, Times, and the 'Right Time'; Chronos and Kairos"

The Monist, Oxford University Press

(1969)

https://www.jstor.org/stable/27902109

12; WEBSITE

"Leap Second Announcements"

International Earth Rotation and Reference Systems Service

https://www.iers.org/IERS/EN/Publications/Bulletins/bulletins.html

LEVEL 2

1; WEBSITE

"May 1801: Thomas Young and the Nature of Light"
American Physical Society
https://www.aps.org/archives/publications/apsnews/200805/physicshistory.cfm

2; LECTURE

Thomas Young
"On the theory of light and colours"
The Royal Society
(1802)
https://royalsocietypublishing.org/doi/10.1098/rstl.1802.0004

3; LECTURE

Thomas Young
"Experiments and calculations relative to physical optics"
The Royal Society
(1804)
https://royalsocietypublishing.org/doi/abs/10.1098/rstl.1804.0001

4; ARTICLE

Peter Rodgers
"The double-slit experiment"
Physics World
(2002)
https://physicsworld.com/a/the-double-slit-experiment/

5; SCIENTIFIC PAPER
P. G. Merli; G. F. Missiroli; G. Pozzi
"On the statistical aspect of electron interference phenomena"
American Journal of Physics
(1976)
https://pubs.aip.org/aapt/ajp/article-abstract/44/3/306/1050404/On-the-statistical-aspect-of-electron-interference

6; WEBSITE
Alain Aspect; John F. Clauser; Anton Zeilinger
"Nobel Prize in Physics 2022 for experiments with entangled photons, establishing the violation of Bell inequalities and pioneering quantum information science"
Nobel Prize Outreach AB
(2022)
https://www.nobelprize.org/prizes/physics/2022/summary/

7; WEBSITE
Erwin Schrödinger
"Schrödinger's Cat"
NASA
(1935)
https://fermi.gsfc.nasa.gov/science/constellations/pages/schrodinger.html

8; SCIENTIFIC PAPER

Albert Einstein; Boris Podolsky; Nathan Rosen

"Can Quantum-Mechanical Description of Physical Reality Be Considered Complete?"

American Physical Society

(1935)

https://journals.aps.org/pr/abstract/10.1103/PhysRev.47.777

9; SCIENTIFIC PAPER

Stuart J. Freedman; John F. Clauser

"Experimental Test of Local Hidden-Variable Theories"

American Physical Society

(1972)

https://journals.aps.org/prl/abstract/10.1103/PhysRevLett.28.938

10; ARTICLE

John G. Cramer

"Quantum Entanglement Across Time"

CENPA

(2019)

https://www.npl.washington.edu/AV/altvw203.html

11; Scientific Paper

E. Megidish; A. Halevy; T. Shacham; T. Dvir; L. Dovrat; H. S. Eisenberg

"Entanglement Between Photons that have Never Coexisted"

American Physical Society

(2013)

https://journals.aps.org/prl/abstract/10.1103/PhysRevLett.110.210403

SIDE-LEVEL 2

1; PDF

"Gateway Intermediate Workbook"
CIA
(1977)
https://www.cia.gov/readingroom/document/cia-rdp96-00788R001700210023-7

2; PDF

"The Gateway Program"
CIA
(1977)
https://www.cia.gov/readingroom/document/cia-rdp96-00788R001700270006-0

3; PDF

"Analysis and Assessment of Gateway Process"
CIA
(1983)
https://www.cia.gov/readingroom/document/cia-rdp96-00788r001700210016-5

LEVEL 3

1; BOOK

David W. Clippinger
"Cultivating Qi"
(2016)

2; BOOK

Eileen Day McKusick
"Electric Body, Electric Health"
(2021)

3; BOOK

Sally Adee
"We Are Electric"
(2024)

4; BOOK

David R. Hawkins
"Power vs. Force"
(1985)

LEVEL 4

1; DICTIONARY

"Intuition"

Oxford English Dictionary

https://www.oed.com/dictionary/intuition_n

2; DICTIONARY

"Intuition"

Cambridge Dictionary

https://dictionary.cambridge.org/dictionary/english/intuition

3; DICTIONARY

"Intuition"

Collins Dictionary

https://www.collinsdictionary.com/dictionary/english/intuition

4; SCIENTIFIC STUDY

Rollin McCraty; Mike Atkinson; Raymond Trevor Bradley

"Electrophysiological Evidence of Intuition: Part 1. The Surprising Role of the Heart"

HeartMath Institute

https://www.heartmath.org/research/research-library/intuition/electrophysiological-evidence-of-intuition-part-1/

5; SCIENTIFIC STUDY

Rollin McCraty; Mike Atkinson; Raymond Trevor Bradley
"Electrophysiological Evidence of Intuition: Part 2. A System-Wide Process?"
HeartMath Institute
https://www.heartmath.org/research/research-library/intuition/electrophysiological-evidence-of-intuition-part-2/

LEVEL 5

1; WEBSITE

Extuity
https://www.extuity.com

LEVEL 6

1; WEBSITE

Emotion-Smart
https://www.emotion-smart.org

LEVEL 7

1; WEBSITE

Chi Fu
https://www.chifu.org

2; BOOK

David W. Clippinger
"Cultivating Qi"
(2016)

3; BOOK

Stephen LaBerge; Howard Rheingold
"Exploring the World of Lucid Dreams"
(1994)

4; BOOK

Charlie Morley
"Dreams of Awakening"
(2024)

5; WEBSITE

Fiona Broome
"The Mandela Effect"
Mandela Effect
https://www.mandelaeffect.com

SIDE-LEVEL 7

1; BOOK

Paul Pearsall
"The Heart's Code"
(1998)

2; PDF

"Cellular Memory in Organ Transplants"
https://www.esalq.usp.br/lepse/imgs/conteudo_thumb/Cellular-Memory-in-Organ-Transplants.pdf

Printed in Great Britain
by Amazon

e644ee5e-e9ec-4eae-b353-8c6adf616623R01